Birkhäuser

Progress in Mathematical Physics
Volume 67

For further volumes:
http://www.birkhauser-science.com/series/4813

Bertrand Duplantier • Vincent Rivasseau
Editors

Henri Poincaré, 1912–2012

Poincaré Seminar 2012

Birkhäuser

Editors
Bertrand Duplantier
Institut de Physique Théorique
CEA Saclay
Gif-sur-Yvette Cedex, France

Vincent Rivasseau
Laboratoire de Physique Théorique
Université Paris-Sud
Orsay, France

ISSN 1544-9998 ISSN 2197-1846 (electronic)
ISBN 978-3-0348-0833-0 ISBN 978-3-0348-0834-7 (eBook)
DOI 10.1007/978-3-0348-0834-7
Springer Basel Heidelberg New York Dordrecht London

Library of Congress Control Number: 2014955371

Mathematics Subject Classification (2010): 01Axx, 00A30, 30F10, 30F35, 37-03, 60-03, 78-03

Cover design: deblik, Berlin

Printed on acid-free paper

Springer Basel is part of Springer Science+Business Media (www.birkhauser-science.com)

Contents

Contributors

Olivier Darrigol
UMR SPHere, CNRS, Université Paris VII, Paris, France

Alain Chenciner
Département de mathématique, Université Paris VII, Paris, France and IMCCE, Paris Obervatory

Laurent Mazliak
Laboratoire de Probabilités et Modéles Aléatoires, Université Pierre et Marie Curie, Paris, France

François Béguin
Laboratoire Analyse, Géometrie et Applications, Université Paris 13 – Sorbonne Paris Cité, Paris, France

Philippe Worms
Vie des Hauts Production, Besançon, France

This thread began with the Sorbonne lectures on the mathematical theories of light, in which the young Poincaré applied his superior mathematical skills to the theory of diffraction, and in which at the same time he developed a structuralist-conventionalist philosophy of physical theory. His subsequent reflections on the foundations of electromagnetism largely depended on this philosophy and on the relation between optics and electromagnetism. He understood that the conceptual difficulties of this theory had to do with the fundamental role of light in the measurement of space and time, thus anticipating Einstein's operational critique of time and simultaneity. His analysis of the optical ether led Poincaré to the principle of relativity in the late 1890s, and to the fully covariant formulation of Lorentz's theory in 1905. Yet, as shown in this learned article, he ended up preserving the ether as the reference for true space and time and as the carrier of all momentum and energy, as a consequence of the same philosophy that had permitted his earlier questioning of the naive mechanical ether.

The second article, the authoritative "Poincaré and the Three-Body Problem" by Alain Chenciner, offers an exquisitely detailed perspective on the monumental work of Poincaré on this subject. It was only incidentally that Poincaré became a mathematical physicist and an astronomer, first holding for a decade the Sorbonne Chair of Calculus of Probability and Mathematical Physics. But it was also a stroke of luck for science. With the author, we follow the steps of Poincaré in the 1889 Memoir, *Sur le problème des trois corps et les équations de la dynamique* (and its corrected version of 1890 in *Acta Mathematica*), which contained the first mathematical description of chaotic behavior in a dynamical system, won him the prize awarded at the occasion of the 60th birthday of King Oscar II of Sweden, and brought him international fame. Then came the extraordinary three volumes of *Les méthodes nouvelles de la mécanique céleste*, published in 1892, 1893 and 1899, which revolutionized celestial mechanics and established the modern theory of dynamical systems. The later works, such as the Poincaré-Birkhoff theorem, appeared in 1912, the very year of Poincare's untimely death.

The author underlines the progression from the search of periodic and quasi-periodic solutions of the three-body problem to the advanced analytic, topological, geometric and probabilistic aspects of the *Méthodes nouvelles*. Highlights are a precise definition of the Poincaré-Lindstedt perturbative series at any order and the breakthrough discovery of their divergence, which led to the modern aspects of the theory (return map, Poincaré's recurrence theorem, ergodic behavior and chaos).

Chenciner takes the time to explain in great detail the mathematics used, and often invented, by a Poincaré inspired by the physical context of his subject, which makes this review extremely readable and valuable. The text is illustrated by beautifully hand-drawn figures and peppered with wide excerpts of citations, usually in French with English translation. They thus allow the reader to enjoy Poincaré's unique and vivid style, while entering into the depth of his foundational work in mechanics.

The thorough study by Laurent Mazliak, "Poincaré's Odds", offers an original and scholarly presentation of the work by Poincaré in probability theory. A challenge in dealing with this topic is that probability penetrated Poincaré's mathematics almost against his intentions, forcing his hand several times. He had indeed received his scientific training in the second half of the 19th century, at a time when Newton's mechanics and Laplace's determinism were the Alpha and Omega of the scientific explanation of the physical world.

The first appearance of the word *probability* in Poincaré's works, in the 1890 probabilistic statement of his famous theorem of recurrence for dynamical systems, involved no intrinsic randomness and was rather a convenient way of expressing the rarity of exceptional non-recurrent trajectories. However, following Boltzmann's and Maxwell's works, statistical mechanics was taking a growingly important role on the scientific stage. After several incisive exchanges with the British physicist Peter Guthrie Tait about Poincaré's 1892 book on Thermodynamics, the latter decided to study and teach the kinetic theory of gases, and he became over the years convinced of the unavoidability of the new approach. During the last twenty years of his life, he devoted theoretical studies, such as the "method of arbitrary functions", as well as philosophical works, such as the chapter on calculus of probabilities in *Science and Hypothesis*, to the question of randomness and its measurement, with a major aim: to identify the situations in which the scientific method required the use of probability theory. A careful examination by Mazliak of these texts, and especially of the two editions of Poincaré's treatise on probability (1896 and 1912) provides us with a clear vision of the evolution of the latter's thinking on the subject. Poincaré seemingly showed little taste for new mathematical techniques, such as measure theory and Lebesgue integration, which could have provided decisive tools to tackle numerous problems.

As a forerunner to its modern development, Poincaré was considered the leading French authority on probability theory. This explains his role during the Dreyfus Affair, when in 1906 he was asked to shed light on the protagonist Bertillon's dubious use of probability concepts. Émile Borel, Poincaré's main direct disciple, developed his approach to a great extent, and influenced the use of probabilities not only in physics, but also in numerous social sciences involving risk and decision. Poincaré's philosophy of randomness also attracted the keen interest of several Czech academics, such as the philosopher Karel Vorovka and the mathematician Bohuslav Hostinský. Partly through them, Poincaré's considerations on card shuffling and ergodicity met a considerable development in the years following 1920, with the emergence of the general theory of Markov chains, which became so fundamental to probability theory in the 20th century.

In "Henri Poincaré and the Uniformization of Riemann Surfaces", François Béguin takes us on a fascinating journey following the maturation of a great mind confronted by an important problem. Through twenty-six years and six succesive versions, the author describes for us the unfolding and deepening of Poincaré's thoughts on the subject up to his ultimate result, which combines a simple and definitive mathematical formulation with an elegant and physical proof. The author thus unveils how uniformization connects together the various facets of Poincaré's work and exemplifies his distinctive fusion of mathematics and physics.

When Poincaré started his work on uniformization, his aim was to understand multi-valued complex analytic functions, and in particular how to parametrize an algebraic curve, as defined by $f(x, y) = 0$, such that both complex variables x and y are single-valued functions of a single other variable z. To this effect, he introduced functions and their invariance groups that he called "Fuchsian". In a famous burst of creativity, he realized that the transformations of his "Fuchsian groups" were identical to those of non-Euclidean (hyperbolic) geometry. This led him to the first two uniformization theorems (1881), which applied first to the Riemann sphere minus a finite number of real points, and then to algebraic Riemann surfaces, modulo a finite number of points. These brilliant steps brought the young Poincaré in contact with Felix Klein for a fruitful scientific correspondence and rivalry, which resulted in their simultaneous statement in 1882 of the "true" uniformization theorem for algebraic Riemann surfaces by means of the so-called continuity method. This was followed in 1883 by the uniformization of (multi-valued) analytic functions, a general result which required the invention by Poincaré of the universal cover of a Riemann surface. The fifth approach of 1898 involved solving the Liouville equation as an alternative method for uniformizing algebraic Riemann surfaces.

Finally, in order to solve Hilbert's 22nd problem, namely to complete the uniformization program for non-algebraic Riemann surfaces, Koebe and Poincaré independently proved in 1907 the modern version of the theorem: *Every simply connected Riemann surface is biholomorphic to the Riemann sphere, the complex plane, or the unit disk.* Their methods were different, and Poincaré used a beautiful, physically inspired, "sweeping method" (i.e., the sweeping of electric charges out of a sequence of disks) which is described in full detail in this article, together with Koebe's simplified proof. Modern mathematical physics, from conformal field theory to string theory and quantum gravity rests on these solid foundations.

In the final chapter, "Harmony and Chaos, On the Figure of Henri Poincaré", PHILIPPE WORMS briefly describes the circumstances and aims of the film that bears the same title, and was projected on the day of the Seminar. Six renowned mathematicians and physicists spent several days of joyful reflection on Poincaré's work in an isolated country house. They speak of geometry, intuition and truth; with this film we embark on a poetical journey unveiling a new mathematical world. The filmmaker successfully focuses, through mathematical scenes and physical experiments, on their emotional relationship to an often elusive truth. It is evoked for us by Poincaré's first words in *La valeur de la science*:

> *"Seeking the truth should be the aim of our activity, it is the only worthy end.* [...] *If we want to gradually free men from material concerns, it is so they can use their newly reacquired freedom to study and gaze at the truth."*

At the end of the same text, we find "harmony" and "beauty":

> *"Mais ce que nous appelons la réalité objective, c'est, en dernière analyse, ce qui est commun à plusieurs êtres pensants, et pourrait être commun à tous ; cette partie commune, comme nous le verrons, ce ne peut être que l'harmonie exprimée par des lois mathématiques.*
>
> *C'est donc cette harmonie qui est la seule réalité objective, la seule vérité que nous puissions atteindre ; et si j'ajoute que l'harmonie universelle du monde est la source de toute beauté, on comprendra quel prix nous devons attacher aux lents et pénibles progrès qui nous la font peu à peu mieux connaître."*

This book, by the breadth of topics covered in Poincaré's legacy, should be of broad interest to mathematicians, physicists and philosophers of science. We further hope that the continued publication of this series of Proceedings will serve the scientific community, at both the professional and graduate levels. We thank the Commissariat à l'Énergie Atomique et aux Énergies Alternatives (Division des Sciences de la Matière), the Daniel Iagolnitzer Foundation, and the École polytechnique for sponsoring this Seminar. Special thanks are due to Chantal Delongeas for the preparation of the manuscript.

Saclay & Orsay
May 2014

Bertrand Duplantier
Institut de
Physique Théorique
Saclay, CEA, France
bduplantier@cea.fr

Vincent Rivasseau
Laboratoire de
Physique Théorique
Université d'Orsay, France
rivass@th.u-psud.fr

Henri Poincaré, 1912–2012, 1–50
Poincaré Seminar 2012

Poincaré's Light

Olivier Darrigol

Abstract. Light is a recurrent theme in Henri Poincaré's mathematical physics. He scrutinized and compared various mechanical and electromagnetic theories of the optical ether. He gave an important boost to the mathematical theory of diffraction. He analyzed the difficulties encountered in applying Lorentz's electromagnetic theory to optical phenomena. He emphasized the metrological function of light in the nascent theory of relativity. Amidst these optical concerns, he derived his philosophy of *rapports vrais*, hypotheses, conventions, and principles, which in turn oriented the critical enterprise through which he pioneered relativity theory.

Reflections on light, its nature, and its propagation pervade Henri Poincaré's works in physics, from his Sorbonne lectures of 1887–88 to his last considerations on geometry, mechanics, and relativity. These reflections enabled him to solve long-standing problems of mathematical optics, most notoriously in diffraction theory; they nourished his criticism of *fin de siècle* electrodynamics; and they inspired a good deal of his philosophy of science. Although Poincaré began to write on celestial mechanics before he did on optics, and although his first physics course at the Sorbonne was on mechanics, he chose the *Théorie mathématique de la lumière* as the topic of the first course he gave from the prestigious chair of Physique mathématique et calcul des probabilités.[1]

Why did Poincaré favor optics over any other domain of physics? (astronomy and mechanics were not regarded as physics proper, at least according to the definition of classes in the French Academy of Sciences). A first hint is found in the obituary that Poincaré wrote for Alfred Cornu, one of his physics professors at the Ecole Polytechnique:

[1]The following abbreviations are used: *ACP, Annales de chimie et de physique; AP, Annalen der Physik; CR*, Académie des sciences, *Comptes rendus hebdomadaires des séances*; *PO*i, Henri Poincaré, *Œuvres*, 11 vols. (Paris, 1954), vol. i; *PRS*, Royal Society of London, *Proceedings.* On Poincaré's early biography, cf. Gaston Darboux, "Eloge historique de Henri Poincaré, membre de l'Académie, lu dans la séance publique annuelle du 15 décembre 1913," in *PO*9, VII–LXXI; Scott Walter, "Henri Poincaré's student notebooks, 1870–1878," *Philosophia scientiae*, 1 (1996), 1–17.

> He has written much on light. Even though he left his mark on every part of physics, optics was his favorite topic. I surmise that what attracted him in the study of light was the relative perfection of this branch of science, which, since Fresnel, seems to share both the impeccable correction and the austere elegance of geometry. In optics better than in any other domain, he could fully satisfy the natural aspiration of his mind for order and clarity.

It seems reasonable to assume that Poincaré was here projecting the reasons for his own predilection. The foreword to his optical lectures begins with: "Optics is the most advanced part of physics; the theory of undulations forms a whole that is most satisfactory to the mind." Being a mathematician with a special fondness for geometric reasoning, Poincaré was naturally drawn to the elegant and powerful geometry of Augustin Fresnel's theory. There also are biographical and cultural reasons for Poincaré's interest in optics. He got his first notions of optics at the lycée of Nancy, and much more at the Ecole Polytechnique, where the two physics professors Jules Jamin and Alfred Cornu and the physics répétiteur Alfred Potier were leading experts in optics. This is no coincidence. In France, elite physicists had long favored optics as a field of research, because it was closely related to the most prestigious sciences of astronomy and mathematics, and because France had excelled in this domain since Fresnel's decisive contributions to the wave theory of light.[2]

From Cornu's course at the Ecole Polytechnique, Poincaré learned the basics of modern optics: the representation of light as a transverse vibration of an elastic medium (the ether); the empirical laws of dispersion; Fresnel's theory of diffraction; Fresnel's construction of rays in anisotropic media; and the finite velocity of light and related phenomena (stellar aberration and the Fresnel drag). Cornu, like most teachers of optics, avoided the deeper theories that involved the nature of the ether and its partial differential equations of motion. There were too many such theories and no evident criterion to select among them; their exposition would have required more advanced mathematics than used in typical physics courses; and the average French physicist of Cornu's time, being first and foremost a sober experimentalist, had little interest in theoretical speculation.[3]

[2]Poincaré, "La vie et les œuvres d'Alfred Cornu," *Journal de l'Ecole Polytechnique*, 10 (1905), 143–155, on 146 [Il a beaucoup écrit sur la lumière; si, en effet, il a laissé sa trace dans toutes les parties de la Physique, c'est surtout pour l'Optique qu'il avait de la prédilection. Je crois que ce qui l'attirait dans l'étude de la lumière, c'est la perfection relative de cette branche de la Science, qui, depuis Fresnel, semble participer à la fois de l'impeccable correction et de la sévère élégance de la Géométrie elle-même. Là, il pouvait, mieux que partout ailleurs, satisfaire pleinement les aspirations naturelles de son esprit d'ordre et de clarté.]; *Leçons sur la théorie mathématique de la lumière, professées pendant le premier semestre* 1887–1888*, rédigées par J. Blondin* (Paris, 1889), I [L'Optique est la partie la plus avancée de la physique; la théorie dite des ondulations forme un ensemble vraiment satisfaisant pour l'esprit.]. On optics and astronomy in 19th-century France, cf. John Davis, "The influence of astronomy on the character of physics in mid-nineteenth century France," *Historical studies in the physical sciences*, 16 (1986), 59–82.
[3]Alfred Cornu, *Cours de Physique, première division,* 1874–1875, cours autographié (Paris: Ecole Polytechnique, 1875).

In contrast, Poincaré judged that his mathematical skills would be best employed if he lectured on the various theories of the ether. He was joining a French tradition of "physique mathématique" in which deeper theory tended to be left to the mathematicians. The challenge was especially high in optics, because of the historical intricacies of its theoretical development. In order to understand the stakes and contents of Poincaré's course, it is necessary to know something about the voluminous literature that Poincaré had to digest and criticize.

1. Optical ether theories

The ether before Poincaré

Although Fresnel's main results, established in the 1820s, can be understood and justified without reference to his precise concept of the ether, there is no doubt that this concept helped him accept the transverse character of luminous vibrations and the kind of elastic response needed to explain propagation in crystals. Fresnel, and most French theorists after him, believed that the ether was a regular lattice of point-like molecules interacting through central forces. They believed the finite spacing of the molecules to be necessary for transverse vibrations. Indeed in a continuous medium ruled by point-to-point central forces, there cannot be any elastic response to a shearing deformation (because the net force between two adjacent layers remains unchanged during a mutual slide of these layers). Moreover, Fresnel believed that the molecular interaction could be adjusted so that his ether would be rigid with regard to optical vibrations and yet liquid with regard to the penetration of ordinary matter. Fresnel's description of his ether was mostly qualitative: he did not derive equations of motion at the molecular or at the medium scale.[4]

The first author to do so was the mathematician Augustin Cauchy, on the basis of the molecular theory of elasticity that Claude Louis Navier, Siméon Denis Poisson, and Cauchy himself had recently developed. For average displacements over volume elements including many molecules, this theory leads to second-order hyperbolic differential equations of motion involving fifteen arbitrary constants in the general, anisotropic case, and allowing for longitudinal vibrations never seen in optics. In order to account for Fresnel's laws of propagation in anisotropic media, which only involve six constants, Cauchy imposed ad hoc conditions on the elasticity constants. In order to account for Fresnel's laws for the intensity of reflected and refracted light at the interface between two homogeneous isotropic transparent media, which only involve transverse vibrations, he adopted boundary conditions incompatible with mechanical common sense (in particular, the displacement of

[4]On Fresnel's theory, cf. Emile Verdet, Introduction, notes, and comments to Augustin Fresnel, *Œuvres complètes d'Augustin Fresnel, publiées par Henri de Sénarmont, Émile Verdet et Léonor Fresnel*, 3 vols. (Paris, 1866–1870); Edmund Whittaker, *A history of the theories of aether and electricity*, vol. 1: The classical theories (London, 1951); Jed Buchwald, *The rise of the wave theory of light: Optical theory and experiment in the early nineteenth century* (Chicago, 1989); Olivier Darrigol, *A history of optics from Greek antiquity to the nineteenth century* (Oxford, 2012).

the medium failed to be continuous at the interface). Later attempts by Franz Neumann, George Green, George Gabriel Stokes, and Gustav Kirchhoff dropped the molecular picture of the ether and replaced it with a continuum approach based on Cauchy's strain and stress tensors (to put it in modern terms). Although these theories encountered similar difficulties, they were regarded as mostly successful. Most physicists in France and abroad were convinced that the ether was some kind of elastic medium obeying the usual laws of mechanics.[5]

Among the many theories of the elastic ether, the "rotational ether" theory and the "labile ether" theory deserve special attention because they did not share the defects of the former theories: they naturally led to the correct number of elastic constants; they naturally excluded longitudinal vibrations; and their boundary conditions were dynamically correct. The Irish mathematician James MacCullagh proposed the rotational ether theory in 1839 on the basis of the Lagrangian density

$$L = \frac{1}{2}\rho\dot{\mathbf{u}}^2 - \frac{1}{2}(\nabla \times \mathbf{u}) \cdot [K](\nabla \times \mathbf{u}), \tag{1}$$

where $\mathbf{u}(\mathbf{r}, t)$ denotes the displacement of the medium at point $\mathbf{r}$ and at time t, ρ the density of the ether, and [K] a symmetric operator determining the elastic response of the ether. MacCullagh proceeded inductively from Fresnel's and others' semi-empirical laws. Although he understood that no ordinary medium had the purely rotational elasticity assumed in his theory, he accepted this feature as an indication that "the constitution of the ether, if it ever would be discovered, will be found to be quite different from any thing that we are in the habit of conceiving, though at the same time simple and very beautiful." MacCullagh's approach was not to please his contemporaries, who did not believe that a Lagrangian offered sufficient mechanical understanding. It is only much later that two other Irishmen, George Francis FitzGerald and Joseph Larmor, realized that MacCullagh's theory contained the basic structure of James Clerk Maxwell's electromagnetic theory of light. Indeed MacCullagh's equation of motion,

$$\rho\ddot{\mathbf{u}} = -\nabla \times [K](\nabla \times \mathbf{u}), \tag{2}$$

has the electromagnetic counterpart

$$\mu\ddot{\mathbf{H}} = -\nabla \times [\epsilon]^{-1}\mathbf{D}, \tag{3}$$

if we take $\mathbf{H} = \dot{\mathbf{u}}$ for the magnetic force field, and $\mathbf{D} = \nabla \times \mathbf{u}$ for Maxwell's displacement, $\mu = \rho$ the magnetic permeability, and $[\epsilon] = [K]^{-1}$ for the dielectric permittivity (which is an operator in anisotropic media).[6]

In the same year 1839, Cauchy sketched his so-called "third theory" of the ether, revived in 1888 in a slightly different form and renamed "labile ether" by William Thomson (Lord Kelvin). In this theory the elastic constants of the

[5]Cf. Whittaker, ref. 4; Darrigol, ref. 4.

[6]James MacCullagh, "An essay towards the dynamical theory of crystalline reflexion and refraction," Royal Irish Academy of Sciences, *Transactions*, 21 (1848, read 9 Dec. 1839), 17–50. Cf. Darrigol, "James MacCullagh's ether: An optical route to Maxwell's equations?" *European physical journal H*, 2 (2010), 133–172.

generic Cauchy–Green theory are adjusted so that the velocity of longitudinal waves (nearly) vanishes. The associated medium is a bit strange: it has negative cubic compressibility; its equilibrium is indifferent to plane compressions (hence the qualification "labile"); and it is best compared to shaving foam. The optical consequences are exactly the same as in MacCullagh's theory, except that the vibrations of the medium occur perpendicularly to the plane of polarization (empirically defined, for instance, by the reflection plane for light polarized by vitreous reflection) whereas in MacCullagh's theory they occur in the plane of polarization. Again, the theory admits an electromagnetic interpretation. Its anisotropic generalization, provided by Richard Glazebrook, rests on the equation of motion

$$[\rho]\ddot{\mathbf{u}} = -K\nabla \times (\nabla \times \mathbf{u}), \tag{4}$$

whose electromagnetic counterpart reads[7]

$$[\epsilon]\ddot{\mathbf{E}} = -\mu^{-1}\nabla \times (\nabla \times \mathbf{E}). \tag{5}$$

if we take $\mathbf{E} = \dot{\mathbf{u}}$, $\mathbf{B} = -\nabla \times \mathbf{u}$, $[\epsilon] = [\rho]$, and $\mu = K^{-1}$.

In general, there were two classes of mechanical ether theories: those for which the vibration belonged to the plane of polarization (Cauchy 1, MacCullagh, Neumann, Green 1, Kirchhoff), and those for which the vibration was perpendicular to this plane (Fresnel, Cauchy 2, Cauchy 3, Green 2, Stokes, Kelvin, Boussinesq). In the first case, the density of the ether is the same in every homogeneous medium; in the second the elastic constant is the same. The second option was by far the most popular for at least three reasons: it implied a more familiar kind of elasticity; it bore the stamp of Fresnel's authority; it permitted a simple interpretation of the Fresnel drag, as we will see in a moment. Yet the first class of theories is better adapted to anisotropic media because an anisotropic elasticity is easier to imagine than an anisotropic density. In the course of the century, there were several attempts to empirically decide between these two options. For instance, Stokes in 1851 and Ludvig Lorenz in 1860 argued that the observed polarization of diffracted light could only be understood if the vibration was in Fresnel's direction. Much later, in 1890, Otto Wiener used his photographic recording of polarized stationary waves near a metallic reflector to decide in favor of Fresnel's choice.[8]

By Wiener' time, physicists were losing interest in such ether-mechanical questions because Heinrich Hertz's production of electromagnetic waves, in 1888, greatly increased the plausibility of Maxwell's electromagnetic theory of light. Maxwell had arrived at this theory in 1865 on the basis of his field-theoretical interpretation of the received laws of electricity and magnetism. In 1855, guided

[7]Augustin Cauchy, "Mémoire sur la polarisation des rayons réfléchis ou réfractés par la surface de séparation de deux corps isophanes et transparents," *CR*, 9 (1939), 676–691; William Thomson (Lord Kelvin), "On the reflexion and refraction of light," *Philosophical magazine*, 26 (1888), 414–425, 500–501. Cf. Whittaker, ref. 4, 145–147; Darrigol, ref. 4, 235–236. Glazebrook justified the anisotropic density $[\rho]$ by analogy with the anisotropic effective mass of a solid ellipsoid immersed in a perfect liquid.

[8]Cf. Whittaker, ref. 4, 328.

propagation velocity independent of the frequency of the waves. In contrast, there were many theories of dispersion based on the mechanical ether. Fresnel originally suggested that the molecular structure of his ether implied a modification of the propagation velocity when the wavelength became comparable to the spacing of the molecules. More precisely, Cauchy showed that the finite spacing of the molecules implied terms of differential order higher than two in the macroscopic equation of propagation. Unfortunately, this simple theory implied that vacuum should itself be dispersive. Subsequent theories by Cauchy and by his followers Charles Briot, Emile Sarrau, and Joseph Boussinesq avoided this and other pitfalls by increasingly separating the ether from the embedded matter. In the end, Boussinesq assumed that the ether and its properties were completely independent of the inclusion of material molecules. After the discovery of anomalous dispersion, in the 1870s Wolfgang Sellmeier and Helmholtz attributed a proper frequency of oscillation to the material molecules and interpreted dispersion as the result of the coupling of the ethereal vibrations with the material oscillators.[13]

This sample of the many ether theories available at the time of Poincaré's lectures should be sufficient to convey the difficulty of conceiving an ether compatible with the known variety of optical phenomena. It also gives an idea of the skills required for the ether builders: a deep understanding of elasticity theory, a firm grasp of the phenomenological laws established by Fresnel and others, an innovative theory of electrodynamics in Maxwell's case, familiarity with the general principles of dynamics in Newtonian and Lagrangian form, and fluency in the calculus of partial differential equations.

Poincaré's first optical lectures (1887–88)

When Poincaré prepared his lectures, Hertz had not yet performed his famous experiments and there was no clear winner among the various theories of the ether. For a pedagogue, the reasonable course would have been to choose among the various competing theories and to dwell on the favorite. This is not what Poincaré did. On the contrary, he expounded no less than eight theories, by Fresnel, Cauchy, Lamé, Briot, Sarrau, Boussinesq, Neumann, and MacCullagh; and he expressed his intention to deal with Maxwell's theory in a subsequent course. This is what physicists occasionally do when they write synthetic reports about the present state of a given domain of physics, as occurred at the British Association for the Advancement of Science and in German encyclopedias in the nineteenth century. But this is not what a good teacher is supposed to do. Poincaré had the excuse of addressing students who had already attended an optics course, and he firmly believed that comparison was the road to truth:[14]

> The theories propounded to explain optical phenomena by the vibrations of an elastic medium are very numerous and equally plausible.

[13]Cf. Darrigol, ref. 4, 244–260. For the sake of brevity, I do not discuss magneto-optics, which also played a role in the selection between various ether theories (and to which Poincaré later contributed a theory of the anomalous Zeeman effect): cf. Buchwald, ref. 10.
[14]Poincaré, ref. 2, II.

> It would be dangerous to confine oneself to one of them. That would bring the risk of blind and therefore deceptive confidence in this one. We must therefore study all of them. Most important, comparison tends to be highly instructive.

Upon reading Poincaré's lectures, one wonders how, in a presumably short time, he could consult so many authors and assimilate so many theories. The answer may perhaps be found in a later remark of his:

> When I read a memoir, I am used to first glance at it quickly so as to get an idea of the whole, and then to return to the points which seem obscure to me. I find it more convenient to redo the demonstrations rather than [going through] those of the author. My demonstrations may be much inferior in general, but for me they have the advantage of being mine.

On the one hand, this method leads to a special clarity, homogeneity, and depth of Poincaré's exposition. On the other hand, it implies departures from the actual contents and intentions of the expounded theories. Not being a historian, Poincaré had more to win on the first account than he had to lose on the second. In particular, the re-demonstration strategy helped him identify shared systems of equations and mathematical structures.[15]

From a mathematical point of view, two distinct ether theories generally differ in two manners: by the partial differential equations of motion and by the boundary conditions at the interface between two media. Poincaré saw that for the most successful theories, the equations of motion in two different theories were related by a simple transformation of the vector representing the vibration. For instance, in MacCullagh's theory the equation of motion is given by equation (2):

$$\rho\ddot{\mathbf{u}} = -\nabla \times [K](\nabla \times \mathbf{u}),$$

Consequently, the vector $\mathbf{v}$ such that

$$\dot{\mathbf{v}} = [K](\nabla \times \mathbf{u}) \tag{8}$$

satisfies

$$[K]^{-1}\ddot{\mathbf{v}} = -\rho^{-1}\nabla \times (\nabla \times \mathbf{v}), \tag{9}$$

in which we recognize the equation of motion (4) of the labile-ether theory if only $[K]^{-1}$ is reinterpreted as a density and ρ^{-1} as an elastic constant. These equations remain valid in heterogeneous media for which the parameters vary in space. Consequently, the boundary conditions at the interface between two homogeneous media can be obtained by taking the limit of a continuous transition layer when the thickness of this layer reaches zero. Thanks to this subterfuge, the mathematical equivalence between the two theories becomes complete. Physical

[15] Poincaré to Mittag-Leffler, 5 Feb. 1889, in *PO*11, p. 69 [Les théories proposées pour expliquer les phénomènes optiques par les vibrations d'un milieu élastique sont très nombreuses et également plausibles. Il serait dangereux de se borner à l'une d'elles; on risquerait ainsi d'éprouver à son endroit une confiance aveugle et par conséquent trompeuse. Il faut donc les étudier toutes et c'est la comparaison qui peut surtout être instructive.]

equivalence follows from the remark that the equations of motion (2) and (9) lead to the same expression of the energy density of the vibration, which gives the luminous intensity.[16]

From this equivalence between the various ether theories, Poincaré concluded that it was impossible to empirically determine the direction of the optical vibration. As we will see in a moment, he refuted the theoretical basis of Stokes's determination of this direction from the polarization of diffracted light. Later, in the 1890s, Poincaré argued that Wiener's stationary-wave experiment equally failed to determine the direction of polarization because in a mechanical theory of the ether there was no reason to assume that the metallic surface reflecting the light in Wiener's device was a nodal surface for the vibrations. All one could assert was that in the electromagnetic theory of light, this surface was a nodal surface for the electric field. Although Cornu and Potier originally supported Wiener, they soon accepted Poincaré's criticism.[17]

Poincaré drew important philosophical lessons form this equivalence between the various theories of light. On the one hand, he emphasized the universality and stability of "the laws of optics and the equations that relate them analytically." This is the first characterization, of what he later called the *rapports vrais* of a theory, that is, relations that are true in any of the competing formulations of a theory and remain approximately true when this theory is replaced by a better one. On the other hand, Poincaré recognized the usefulness of "doctrines coordinating the equations of the theory," doctrines implying what he later called "indifferent hypotheses." For the optical theories, a first indifferent hypothesis is the choice of the direction of vibration. A second is the molecular versus continuum description of the ether. In his lectures of 1887–88 Poincaré adopted the molecular hypothesis, with the following explanation:

> The theory of undulations rests on a molecular hypothesis. For those who believe to be thus unveiling the cause of the law, this is an advantage; for the others, this is a reason for being suspicious. This suspicion, however, seems to me as little justified as the illusion of the believers. The hypotheses play only a secondary role. I could have avoided them; I did not because the clarity of the exposition would have suffered from it. This is the only reason. Indeed the only things I borrow from molecular hypotheses are the principle of energy conservation and the linear form of the equations, which is the general law of small movements and of all small variations.

To better prove this point, in the 1891–92 sequel to his lectures Poincaré adopted the continuum approach in which the elastic ether is described by Cauchy's strain

---

[16] Poincaré, ref. 2, 399–400. Ludvig Lorenz invented the transition-layer approach.

[17] Otto Wiener, "Stehende Lichtwellen und die Schwingungsrichtung polarisirten Lichtes," *AP*, 38 (1890), 203–243; Poincaré, "Sur l'expérience de Wiener," *CR*, 112 (1891), 325–329; "Sur la réflexion métallique," *CR*, 112 (1891), 456–459. Cf. Scott Walter (ed.), *La correspondance entre Henri Poincaré et les physiciens, chimistes et ingénieurs* (Basel, 2007), 107–108.

and stress tensors. In the foreword to his lectures of 1887–88 he pushed agnosticism so far as to question the reality of the ether:

> It matters little whether the ether really exists; that is the affair of the metaphysicians. The essential thing for us is that everything happens as if it existed, and that this hypothesis is convenient for the explanation of phenomena. After all, have we any other reason to believe in the existence of material objects? That, too, is only a convenient hypothesis; only this will never cease to be so, whereas probably the ether will some day be thrown aside as useless. On this very day, however, the laws of optics and the equations that express them analytically will remain true, at least in a first approximation. It will therefore be always useful to study a doctrine that interconnects all these equations.

In sum, for Poincaré physical hypotheses should not be taken too seriously, although they are useful for the sake of clarification and illustration. In many cases, they only are concrete means to satisfy general principles that have more direct empirical significance, for instance the energy principle or the superposition principle to which Poincaré refers in his defense of the molecular ether. As we will see in a moment, Poincaré later came to favor a more direct application of the principles.[18]

[18]Poincaré, ref. 4, II, III [La théorie des ondulations repose sur une hypothèse moléculaire ; pour les uns, qui croient découvrir ainsi la cause sous la loi, c'est un avantage; pour les autres, c'est une raison de méfiance ; mais cette méfiance me paraît aussi peu justifiée que l'illusion des premiers. Ces hypothèses ne jouent qu'un rôle secondaire. J'aurais pu les sacrifier ; je ne l'ai pas fait parce que l'exposition y aurait perdu en clarté, mais cette raison seule m'en a empêché. En effet je n'emprunte aux hypothèses moléculaires que deux choses: le principe de la conservation de l'énergie et la forme linéaire des équations qui est la loi générale des petits mouvements, comme de toutes les petites variations.], I–II [Peu nous importe que l'éther existe réellement; c'est l'affaire des métaphysiciens ; l'essentiel pour nous c'est que tout se passe comme s'il existait et que cette hypothèse est commode pour l'explication des phénomènes. Après tout, avons-nous d'autre raison de croire à l'existence des objets matériels? Ce n'est là aussi qu'une hypothèse commode ; seulement elle ne cessera jamais de l'être, tandis qu'un jour viendra sans doute où l'éther sera rejeté comme inutile. Mais ce jour-là même, les lois de l'optique et les équations qui les traduisent analytiquement resteront vraies, au moins comme première approximation. Il sera donc toujours utile d'étudier une doctrine qui relie entre elles toutes ces équations.]. On Poincaré's hypotheses and "rapports vrais," cf. David Stump, "Henri Poincaré's philosophy of science," *Studies in history and philosophy of science*, 20 (1989), 335–363; Igor Ly, *Géométrie et physique dans l'œuvre de Henri Poincaré*, Thèse, Université Nancy 2 (2007); João Principe da Silva, "Sources et nature de la philosophie de la physique d'Henri Poincaré," *Philosophia scientiae*, 16 (2012), 197–222; Darrigol, "Diversité et harmonie de la physique mathématique dans les préfaces de Henri Poincaré," in Jean-Claude Pont et al. (eds.), *Pour comprendre le XIX*e *: Histoire et philosophie des sciences à la fin du siècle* (Florence, 2007), 221–240.

2. Diffraction theory

The Kirchhoff–Poincaré approximation

Besides clarifying the contents and interrelations of the various ether theories, Poincaré's optical lectures brought important insights into the theory of diffraction. Not only Poincaré independently recovered Kirchhoff's main results in this domain, but he addressed mathematical difficulties of which Kirchhoff was unaware, he was able to determine the physical conditions under which Kirchhoff's diffraction formula yields correct results, and he pioneered the study of cases of diffraction in which these conditions did not hold.

Fresnel's theory of diffraction is based on the intuitive idea that the vibration at a point situated beyond a diffracting screen is equal to the sum of vibrations emanating from every point of the screen's opening, with an original amplitude and phase equal to those of the vibration that the source would produce in absence of the screen. While Fresnel did not doubt the mechanical soundness of this intuition, mathematicians like Poisson regarded it as unfounded. In the course of time it nonetheless became clear that the formula gave correct predictions in most cases of diffraction. In a bulky memoir of 1851, Stokes obtained a more precise diffraction formula than Fresnel for the polarized transverse vibrations of an elastic solid representing the ether. His derivation was based on an exact representation of the vibration from an unscreened point source as a retarded integral on a plane. In order to obtain the light diffracted by a screen in this plane, Stokes simply restricted the integration to the screen's opening. Efficient though it is, this derivation has two major defects: the surface integral representation of the vibration is not unique, and the truncation of the integral rests on two unwarranted assumptions: that the vibration in the opening of the screen is the same as if the screen were not there, and that the vibration on the unexposed side of the screen strictly vanishes.[19]

In 1882, Kirchhoff removed the first difficulty by relying on a generalization of Green's theorem that Helmholtz had given in an influential memoir on the vibrations of open organ pipes. For monochromatic sound waves of frequency kc, the wave equation has the form

$$\Delta u + k^2 u = 0. \tag{10}$$

It admits the Green functions

$$G_{\mathrm{M}}(\mathbf{r}) = -\frac{\mathrm{e}^{\mathrm{i}k|\mathbf{r}-\mathbf{r}_{\mathrm{M}}|}}{4\pi|\mathbf{r}-\mathbf{r}_{\mathrm{M}}|} \tag{11}$$

such that

$$\Delta G_{\mathrm{M}} + k^2 G_{\mathrm{M}} = \delta(\mathbf{r}-\mathbf{r}_{\mathrm{M}}). \tag{12}$$

For any two functions f and g of $\mathbf{r}$, we have

$$f\Delta g - g\Delta f = \nabla\cdot(f\nabla g - g\nabla f), \tag{13}$$

[19] Gabriel Stokes, "On the dynamical theory of diffraction," Cambridge Philosophical Society, *Transactions*, 9 (1851, read 26 Nov. 1849), 1–62. Cf. Darrigol, ref. 4, 273–276.

whence follows Green's theorem for the volume V delimited by the closed surface ∂V:

$$\int_{\mathrm{V}}(f\Delta g - g\Delta f)\mathrm{d}\tau = \int_{\partial\mathrm{V}}(f\nabla g - g\nabla f)\cdot\mathrm{d}\mathbf{S}, \tag{14}$$

or else

$$\int_{\mathrm{V}}[f(\Delta + k^2)g - g(\Delta + k^2)f]\mathrm{d}\tau = \int_{\partial\mathrm{V}}(f\nabla g - g\nabla f)\cdot\mathrm{d}\mathbf{S}\ . \tag{15}$$

Helmholtz specialized this identity to $f = u$ and $g = G_{\mathrm{M}}$, where u satisfies the free wave equation (10) within the volume V. Call S the boundary ∂V of this volume, V$'$ the complementary volume (on the other side of S), and $H_{\mathrm{M}}(u,\mathrm{S})$ the Helmholtz integral defined by

$$H_{\mathrm{M}}(u,\mathrm{S}) = \int_{\mathrm{S}}(G_{\mathrm{M}}\nabla u - u\nabla G_{\mathrm{M}})\cdot\mathrm{d}\mathbf{S}\ . \tag{16}$$

Then equation (15) implies the following Helmholtz identities:[20]

$$\text{If } \mathrm{M}\in\mathrm{V},\qquad H_{\mathrm{M}}(u,\mathrm{S}) = u(\mathrm{M}) \tag{H}$$

$$\text{If } \mathrm{M}\in\mathrm{V}',\qquad H_{\mathrm{M}}(u,\mathrm{S}) = 0\ . \tag{H$'$}$$

Now consider Kirchhoff's diffracting device of Fig. 1, and apply the Helmholtz identity H to the volume V delimited by the closed surface $s\cup s''$ and by a very large sphere s_∞ containing the whole setup. For any point M within this volume, we have

$$u(\mathrm{M}) = H_{\mathrm{M}}(u,\mathrm{s}) + H_{\mathrm{M}}(u,\mathrm{s}'') + H_{\mathrm{M}}(u,\mathrm{s}_\infty). \tag{17}$$

The third term can be ignored because the vibration never reaches a sufficiently remote surface.[21] Kirchhoff further assumes that for a perfectly black screen

(i) $u = 0$, $\partial u/\partial n = 0$, on the external surface s$''$,
(ii) $u = u_1$, $\partial u/\partial n = \partial u_1/\partial n$ on the internal surface s$'$
(iii) $u = u_1$, $\partial u/\partial n = \partial u_1/\partial n$ on the surface s of the opening,

wherein $\partial/\partial n$ denotes the normal derivative and u_1 the wave created by the source 1 in the absence of the screen ($u_1(\mathbf{r}) \propto \mathrm{e}^{\mathrm{i}k|\mathbf{r}-\mathbf{r}_1|}/|\mathbf{r}-\mathbf{r}_1|$). Under the assumptions (i) and (iii), equation (17) leads to the Kirchhoff diffraction formula

$$u(\mathrm{M}) = H_{\mathrm{M}}(u_1,\mathrm{s}). \tag{18}$$

In his lectures of 1887–88, Poincaré obtained the same formula by similar means, although he was unaware of Helmholtz's and Kirchhoff's memoirs. Unlike Kirchhoff, he realized that the assumptions (i) and (iii) were mutually incompatible. If both assumptions were true, Poincaré reasoned, for the same integration

[20] Gustav Kirchhoff, "Zur Theorie der Lichtstrahlen," Akademie der Wissenschaften zu Berlin, mathematisch-physikalische Klasse, *Sitzungsberichte*, 2 (1882), 641–692; Hermann Helmholtz, "Theorie der Luftschwingungen in Röhren mit offenen Enden," *Journal für die reine und angewandte Mathematik*, 57 (1859), 1–72.

[21] This consideration requires Kirchhoff's extension of the Helmholtz–Green theorem to non-periodic perturbations of the medium, which I omit for the sake of simplicity. I ignore the transverse vector character of the optical vibrations, which does not affect the main results.

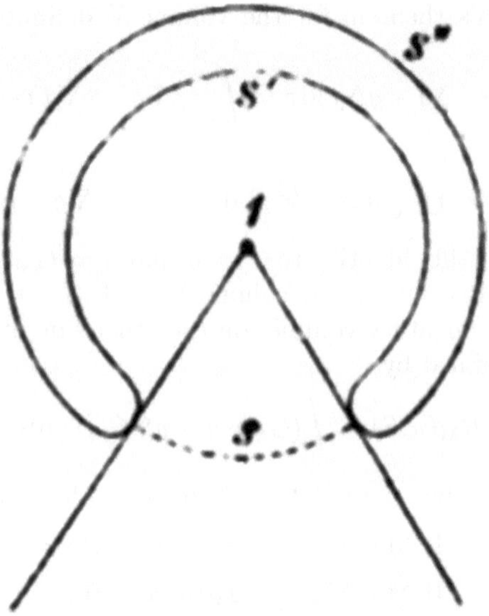

FIGURE 1. Kirchhoff's diffraction problem. The point source 1 is included in the crescent-shaped cavity with fully absorbing walls $s' \cup s''$ and opening s. The observation point is outside the cavity. From Kirchhoff, ref. 20, 80.

surface $s \cup s''$ and for a point M contained in the volume V′ within the closed surface, the Helmholtz identity H′ would yield $H_{\mathrm{M}}(u_1, s) = 0$, which is generally untrue (consider for instance the case when the width of s is a fraction of a wavelength). Even worse, the seemingly natural conditions (i) and (ii) are also incompatible, because the Helmholtz identity H′ for a point M within (in the substance of) the screen and for the volume V delimited by the surfaces $s' \cup s''$, s_∞, and a small sphere s_1 centered on point 1 would then yield

$$H_{\mathrm{M}}(u_1, \mathrm{s}') = -H_{\mathrm{M}}(u_1, \mathrm{s}_1) = u_1 \; ; \tag{19}$$

in the absence of the screen, the same identity (applied to the same surfaces and volume) yields

$$H_{\mathrm{M}}(u_1, \mathrm{s}') + H_{\mathrm{M}}(u_1, \mathrm{s}'') = u_1 \; ; \tag{20}$$

hence the integral $H_{\mathrm{M}}(u_1, \mathrm{s}'')$ would vanish for any surface s″, which cannot be true. Unknown to Kirchhoff and to Poincaré, in 1869 the mathematician Heinrich Weber had proved that for a function satisfying the generalization (10) of Laplace's equation, the boundary condition $u = 0$, $\partial u/\partial n = 0$ on any finite portion of a surface implies that the function u should vanish in the whole (connected) domain in which this equation holds. In order to avoid this paradox, the part of the screen invisible from the source must be slightly illuminated, as can be verified experimentally.[22]

Poincaré not only detected a fundamental inconsistency in his and Kirchhoff's theory, but he showed how to circumvent it. As we just saw, Kirchhoff's assumptions lead to $H_{\mathrm{M}}(u_1, \mathrm{s}) = 0$ for any point M within the surface $\mathrm{s} \cup \mathrm{s}''$. Although the integral $H_{\mathrm{M}}(u_1, \mathrm{s})$ generally differs from zero, it is approximately zero when the width of the opening s largely exceeds the wavelength and when M is not too close to the rim of s. Poincaré further noticed that the approximate vanishing of this integral implied the approximate validity of Kirchhoff's diffraction formula (18) for points M outside the surface $\mathrm{s} \cup \mathrm{s}''$, far enough from its rim and not too deeply within the geometric shadow of the screen. Indeed the integral $H_{\mathrm{M}}(u_1, \mathrm{s})$ and its normal derivative are continuous when the point M crosses the surface s'', and they suddenly increase by u_1 and by $\partial u_1/\partial n$ when M crosses the surface s from within (this property is analogous to the discontinuity of the electric field created by a surface charge). Hence, if the integral $H_{\mathrm{M}}(u_1, \mathrm{s})$ approximately vanishes within $\mathrm{s} \cup \mathrm{s}''$, the values of this integral and of its normal derivative are approximately zero on the exterior side of s'' and they are approximately u_1 and $\partial u_1/\partial n$ on the exterior side of s. In addition, this integral is an exact solution of the Helmholtz equation (10) outside the surface $\mathrm{s} \cup \mathrm{s}''$. Consequently, the function that takes the value u_1 within $\mathrm{s} \cup \mathrm{s}'$ and $H_{\mathrm{M}}(u_1, \mathrm{s})$ outside $\mathrm{s} \cup \mathrm{s}''$ is an approximate solution of the wave equation that approximately meets Kirchhoff's boundary conditions. Poincaré completed this reasoning by evaluating $H_{\mathrm{M}}(u_1, \mathrm{s})$ for small wavelengths and showing that in usual cases of diffraction the Kirchhoff integral provided a good approximation of the distribution of diffracted light. Unfortunately, this remarkable explanation of the otherwise surprising success of Kirchhoff's theory seems to have been forgotten.[23]

Large-angle diffraction

Poincaré was aware of a significant exception to Kirchhoff's approximation: the large-angle diffraction experiments performed by Stokes and a few others. In this case, Stokes's theory predicted that diffraction privileged vibrations perpendicular to the diffraction plane (that is, the plane containing the incoming ray and the diffracted ray). Experimenting with a grating, he found the diffracted light to be polarized in the diffraction plane, thus confirming Fresnel's choice of the

---

[22]Poincaré, ref. 2, 99–118; *Théorie mathématique de la lumière. II. Nouvelles études sur la diffraction. Théorie de la dispersion de Helmholtz* [1st semester 1891–1892], eds. M. Lamotte and D. Hurmuzescu (Paris, 1893), 182–188. For Poincaré being unaware of Kirchhoff's memoir, see ibid, introduction (2 Dec. 1888), on IV: "Dans le chapitre relatif à la diffraction, j'ai développé des idées que je croyais nouvelles. Je n'ai pas nommé Kirchhoff dont le nom aurait dû être cité à chaque ligne. Il est encore temps de réparer cet oubli involontaire; je m'empresse de le faire en renvoyant aux Sitzungsberichte de l'Académie de Berlin (1882...)." On Weber's theorem and diffraction, cf. Arnold Sommerfeld, *Vorlesungen über die theoretische Physik. Band IV: Optik* (Leipzig, 1950), 202.

[23]Poincaré, ref. 2, 115–118 (general reasoning), 118–130 (evaluation of $H_{\mathrm{M}}(u_1, \mathrm{s})$ in the case of a spherical screen with a spherical hole).

direction of the vibration. In 1856, the Stuttgart Professor Carl Holtzmann confirmed Neumann's opposite choice in similar experiments. In 1861, independently of this controversy, Fizeau showed that light diffracted by extremely thin stripes on a metallic surface or by an extremely thin slit was almost completely polarized at large diffraction angles. He explained this result by interference and reflection-based phase shift: in the slit experiments, direct (diffracted) light interferes with light reflected by the edges of the slit, in a different manner for the components of the incoming vibration perpendicular and parallel to the length of the slit because these two components undergo different phase shifts by reflection.[24] In 1886, knowing that earlier experiments on polarization by diffraction had given conflicting results, the Lyon-based physicist Louis Georges Gouy studied the pure case of diffraction by a razor-sharp (metallic) edge. In order to get an observable amount of diffracted light at large angles, he concentrated the light from the sun or from an arc lamp on a point of the edge, and observed the diffracted light through a microscope focused on the edge (Fig. 2). He found that at large angle the diffracted light depended on the polarization of the incoming light, on the material of the edge, and on its sharpness – all against Fresnel's theory; for initially unpolarized light, the internally diffracted light was polarized perpendicularly to the diffraction plane, almost completely so when the angle took it maximal value. Like Fizeau, Gouy surmised that the phenomenon had similarity with metallic reflection, and that the reflectivity of the material and its superficial conductivity played a role when the light traveled near the edge.[25] At the very end of his lectures of 1887–88, Poincaré summarized these results and noted that the conditions for the validity of Kirchhoff's approximation were no longer met. The true boundary conditions at the surface of the diffracting screen had to be taken into account, so that, pace Stokes, none of these experiments could decide between Fresnel's and Neumann's choice of the direction of vibration:[26]

> This disagreement with Mr. Gouy's experiments should not surprise us, for we have said that it was impossible to find a solution of the equation $\Delta\xi + \alpha^2\xi = 0$ that satisfied *exactly* the conditions of the problem. Only by *approximately* satisfying these conditions could we build a theory of diffraction. The approximation was largely sufficient in the usual conditions of diffraction experiments because the neglected quantities are

[24]Stokes, ref. 19; Carl Holtzmann, "Das polarisirte Licht schwingt in der Polarisationsebene," *AP*, 99 (1856), 446–451; Hippolyte Fizeau, "Recherches sur plusieurs phénomènes relatifs à la polarisation de la lumière," *CR*, 52 (1861), 267–278, 1221–1232.

[25]Louis Georges Gouy, "Recherches expérimentales sur la diffraction," *ACP*, 52 (1886), 145–192.

[26]Poincaré, ref. 2, 401 [Ce désaccord entre les expériences de M. Gouy et la théorie de Fresnel ne doit pas nous surprendre, car nous avons dit qu'il était impossible de trouver une solution de l'équation $\Delta\xi + \alpha^2\xi = 0$ satisfaisant *exactement* aux conditions du problème. Ce n'est qu'en y satisfaisant *approximativement* que nous avons pu édifier une théorie de la diffraction. L'approximation était très largement suffisante dans les conditions habituelles des expériences de diffraction ; car les quantités négligées sont alors extrêmement petites. Il n'en est plus de même dans les conditions où M. Gouy s'était placé.]. See also Poincaré, ref. 22 (1893), 195, 213–223.

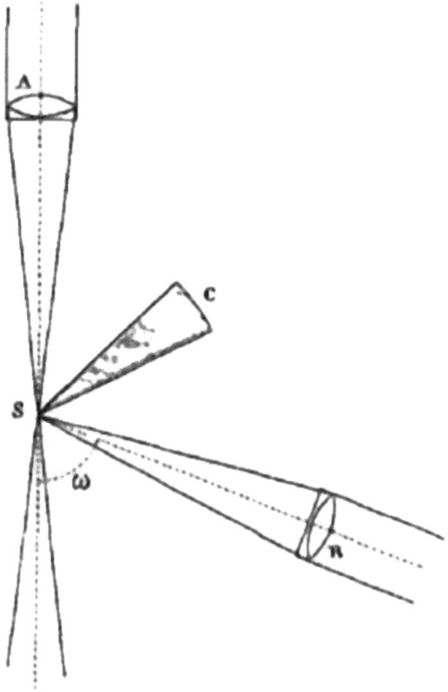

FIGURE 2. Diffraction of convergent light (from the lense A) by the edge (S) of a razor (CS). The diffracted light is observed at angle through a microscope of objective lens R. From Gouy, ref. 25, 148.

extremely small in this case. This ceases to be true in the conditions in which Mr. Gouy operated.

Poincaré returned to this question in 1892 in an attempt to explain the results that Gouy had obtained in his seductively simple device. Poincaré first gave an exact solution to the further simplified problem in which the waves converging on the edge and diverging from it are cylindrical waves (so that the problem becomes bidimensional), the metal of the edge is regarded as a perfect conductor (so that the electric lines of force are perpendicular to the surface of the metal), the edge is perfectly sharp and its angle is infinitely small. The fundamental equations of the problem are the Maxwell equations with the boundary condition that the tangential component of the electric field should vanish on the surface of the metal blade. A development of the fields into Bessel functions and some algebra lead to the following expression for the large-distance amplitude $A_{/\!/}(\omega)$ of the light diffracted at the angle ω, when the incident beam is homogeneous, comprised

> These considerations should be sufficient to show the weak point of Mr. MacDonald's reasoning. It would be important to resume the calculations in a manner that takes this difficulty into account, for we want to know whether the results obtained by Mr. Marconi can be explained by present theories and simply result from the exquisite sensibility of his coheror, or instead prove that the waves are reflected by the upper layers of the atmosphere, these layers being made conductors by their extreme rarefaction.

Poincaré returned to this problem in a series of conferences he delivered at the Ecole Supérieure des Postes et Télégraphes in May–June 1908. There he offered a simple intuitive argument leading to the exponential decay of the intensity of the light diffracted along the curved surface of the earth. In the absence of diffraction, Poincaré reasoned, the radiation emitted by the antenna OC in Fig. 3 would be restricted to the right angle COF, OF being the tangent to the earth sphere at point O. Call I the intensity of the radiation emitted in a small angle FOD above OF. Owing to diffraction, a fraction α of this radiation should be found in the equal angle FHG under OF. Similarly, a fraction of the latter radiation should be found in the equal angle GKL under GH (the fraction is the same because intuitively diffraction into the shadow only depends on the intensity of light at the limit of the shadow). After n iterations, the intensity in the last equal angle is $I\alpha^n$, so that the radiation decreases exponentially with the distance from the antenna.[31]

Later in the same year Poincaré performed a more serious calculation based on the Legendre-polynomial series solution of a Fredholm equation he deduced from the boundary condition on the sphere and from Maxwell's equations. With a few approximations, he obtained a diffracted intensity (more exactly, a surface current) proportional to $\lambda^{1/4}$ if λ denotes the wavelength. In a Göttingen lecture of April 1909 he concluded: "In this manner we can explain the astonishing fact that it is possible, by means of the Hertzian waves of wireless telegraphy, to communicate from the European continent to America, for example." Alas, Poincaré soon detected an error in his asymptotic estimate of the Bessel functions of his spherical-harmonic development: he had neglected some terms which in reality cancelled most of the contribution he had retained. After correcting this error, he found that the diffracted intensity varied as $\mathrm{e}^{-\alpha\theta R^3/\lambda^3}$, wherein θ is the angular distance between the emitter and the receiver, λ the wavelength, R the radius of

supérieures de l'atmosphère rendues conductrices par leur extrême raréfaction.]; Hector Munro MacDonald, "The bending of electric waves round a conducting obstacle," *PRS*, 71 (1903), 251–258; Lord Rayleigh, "On the bending of waves around a spherical obstacle," *PRS*, 72 (1904), 40–41; Poincaré, "Sur la diffraction des ondes électriques: A propos d'un article de M. MacDonald," *PRS*, 72 (1904), 42–52, on 42, 52.

[31]Poincaré, "Conférences sur la télégraphie sans fil," *La lumière électrique*, 4 (1908), 259–266, 291–297, 323–327, 355–359, 387–393, on 323. Cf. Jean-Marc Ginoux, "Les conférences 'oubliées' d'Henri Poincaré: les cycles limites de 1908," http://bibnum.education.fr/files/Poincare-analyse.pdf (last accessed Oct. 2012).

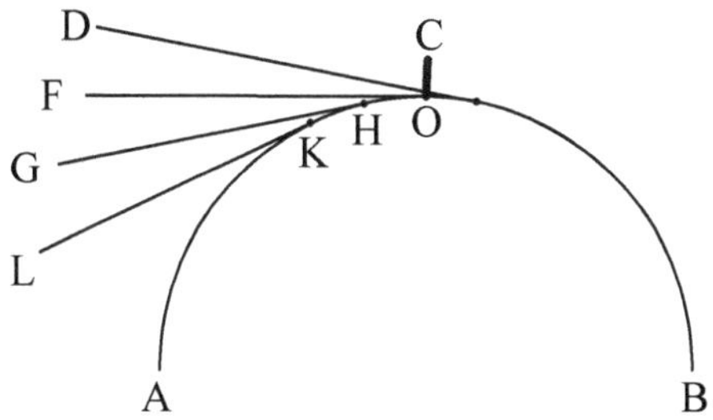

FIGURE 3. Poincaré's diagram for the terrestrial diffraction of Hertzian waves. The waves are emitted by the antenna OC; AOB is the trace of the earth's surface; the lines ending in D, F, G, L are successive tangents to this surface at equidistant points. From Poincaré, ref. 31, 323 (redrawn).

the earth, and α a numerical constant. In his final memoir on this topic, he produced the full corrected calculations with an apology for his "successive palinodes." He now doubted that diffraction could explain long-distance transmission and he noted that the alternative explanation by ionization of the upper atmosphere might perhaps account for the superior quality of nighttime transmissions.[32]

In 1911, an American student of Sommerfeld, Herman William March, filed a dissertation on the propagation of telegraphic waves around the earth and obtained results contradicting Poincaré's. In a letter to Sommerfeld and in a note to the *Compte rendus*, Poincaré identified a fatal error in March's calculation. He also noted that a Trinity Wrangler, John William Nicholson, had confirmed his own exponential law. In fact, Nicholson had improved on Poincaré's method and obtained an estimate (0.696) for the numerical coefficient α in Poincaré's law. As Poincaré noted, this estimate conflicted with recent measurements of long-distance attenuation by Louis Austin's team at the U.S. Naval Wireless Telegraphic Laboratory:

> The attenuation coefficient had been found, *even at daytime*, to be hundred times smaller than the theoretical coefficient resulting from my calculation. The ordinary theory therefore does not account for the facts; something remains to be found.

[32]Poincaré, "Anwendung der Integralgleichungen auf Hertzsche Wellen," in *Sechs Vorträge über ausgewählte Gegenstände aus der reinen Mathematik und mathematischen Physik* (Leipzig, 1910), 23–31, on 31 [Auf diese Weise wird die zunächst staunenerregende Tatsache verständlich, dass es mit Hilfe der in der drahtlosen Telegraphie verwendeten Hertzschen Wellen gelingt, vom europäischen Kontinent z.B. bis nach Amerika zu telegraphieren.]; "Sur la diffraction des ondes hertziennes," *CR*, 149 (1909), 92–93 (error corrected); "Sur la diffraction des ondes hertziennes," *Rendiconti del Circolo Matematico di Palermo*, 29 (1910), 159–269, on 268–269. Nicholson (ref. 33) spotted Poincaré's error independently of Poincaré.

The diffraction theory nonetheless survived Poincaré's death, until in 1918 the Cambridge-trained mathematician George Neville Watson invented the powerful "Watson transformation" that is now used to solve this kind of problem.[33]

3. The nature of white light

The last and least successful intervention of Poincaré in optics *stricto sensu* occurred in a polemic with a physicist he admired particularly, the aforementioned Georges Gouy. In the 1880s Gouy had argued, against Cornu, that the velocity of light in dispersive media as measured by Fizeau's method of the toothed wheel was what we now call the group velocity, not the phase velocity. This interest in chopped or modulated waves brought him to discuss the received view of white light as a random mixture of wave trains of various lengths, origins, and frequencies (as one would expect if the source is made of randomly excited vibrators). In an influential memoir of 1886, Gouy argued that whatever be the detailed mechanism of the production of light, the ethereal motion $s(t)$ on a plane far from the source before entering an optical system could always be represented by a Fourier integral

$$s(t) = \int \tilde{s}(\omega)\mathrm{e}^{\mathrm{i}\omega t}\mathrm{d}\omega \tag{23}$$

and that the time-averaged illumination at the exit of the optical system could be obtained by superposing the illuminations caused by incoming plane monochromatic waves $\mathrm{e}^{\mathrm{i}(\omega t-\mathbf{k}\cdot\mathbf{r})}$ with the weights $|\tilde{s}(\omega)|^2$. Although this result is a trivial consequence of the linearity of the equations of propagation and of Perceval's theorem for Fourier's transforms, it has the counterintuitive consequence that no optical experiment can decide between concepts of natural light that lead to the same frequency distribution $|\tilde{s}(\omega)|^2$. In particular, there should be no way to decide whether the disappearance of fringes in an interference device with large path difference is due to the "complexity" (varying frequency) or to the "irregularity" (disrupted wave trains) of the incoming light.[34]

As Gouy knew, in 1845 Fizeau and Léon Foucault believed to have excluded the second alternative. They used a spectrometer to analyze the light issuing from a double-ray interference device (Fresnel's mirrors) fed by white light. At a given

[33]Herman William March, "Über die Ausbreitung der Wellen der drahtlosen Telegraphie auf der Erdkugel," *AP*, 37 (1912), 29–50; Poincaré to Sommerfeld (c. March 2012), in Walter, ref. 17, 343–344; John William Nicholson, "On the bending of electric waves round the earth," *Philosophical magazine*, 19 (1910), 276–278, 435–437, 757–760; "On the bending of electric waves round a large sphere," *Philosophical magazine*, 19 (1910), 516–537; 20 (1910), 157–172; 21 (1911), 62–68, 281–295; Poincaré, "Sur la diffraction des ondes hertziennes," *CR*, 154 (1912), 795–797, on 797 [Le coefficient d'affaiblissement a été trouvé, même de jour, cent fois plus faible que le coefficient théorique résultant de mon calcul. La théorie ordinaire ne rend donc pas compte des faits, il y a quelque chose à trouver.]. Cf. Gérard Petiau, commentary in *PO*10, 217–219; Yeang, ref. 29, on 398.

[34]Gouy, "Sur le mouvement lumineux," *Journal de physique*, 5 (1886), 354–362. Cf. André Chappert, *L'édification au XIXᵉ siècle d'une science du phénomène lumineux* (Paris, 2004), 247–251.

point of the zone of interference, the spectral component of pulsation ω_0 has an amplitude proportional to $1 + \mathrm{e}^{\mathrm{i}\omega_0\tau}$, wherein τ denotes the delay caused by the path difference. Therefore, the observed spectrum has periodic dark lines whose number increases with the path difference (*spectre cannelé* or band spectrum). Fizeau and Foucault were able to distinguish these lines for path differences as large as seven thousand wavelengths. They concluded:

> The very restricted limits of path difference beyond which one could not [heretofore] produce the mutual influence of two rays depended only on the complexity of light. By using the simplest light that one might obtain, these limits are considerably shifted. – The existence of these phenomena of the mutual influence of two rays in the case of a large path difference is interesting for the theory of light, for it reveals in the emission of successive waves a persistent regularity that no phenomenon earlier suggested.

Gouy flatly rejected this conclusion, since in his view any irregularity in the source was equivalent to a spread in the Fourier spectrum of the vibration:[35]

> Thus, the existence of interference fringes for large path differences does not at all imply the regularity of the incoming luminous motion. This regularity exists in the spectrum, but it is the spectral apparatus that produces it by separating more or less completely the various simple motions which heretofore only had a purely analytical existence.

Poincaré took Fizeau's defense in a note of 1895 for the *Comptes rendus* (Fizeau died the following year). Poincaré first argued that Gouy's reasoning led to the absurd consequence that a source of light, when seen through a spectroscope, should appear permanently illuminated even if the source was turned off. Indeed according to Gouy the spectroscope separates the Fourier components of the vibration, and by definition Fourier components do not depend on time. In order to avoid this paradox, Poincaré introduced the finite resolution of the spectroscope. The amplitude of vibrations at a point of the interference zone can the be written as

$$a(t) \propto \int (1 + \mathrm{e}^{\mathrm{i}\omega\tau})\chi(\omega - \omega_0)\tilde{s}(\omega)\mathrm{e}^{\mathrm{i}\omega t}\mathrm{d}\omega, \tag{24}$$

wherein $\chi(\omega - \omega_0)$ characterizes the frequency selection by the spectroscope. In the case of infinite resolution, $\chi(\omega - \omega_0) = \delta(\omega - \omega_0)$, so that

$$a(t) \propto \mathrm{e}^{\mathrm{i}\omega_0 t}\tilde{s}(\omega_0)(1 + \mathrm{e}^{\mathrm{i}\omega_0\tau}) \tag{25}$$

and the corresponding intensity does not depend on time. In reality, the resolution of the spectroscope is limited by its finite aperture, and the characteristic function

---

[35] Fizeau and Léon Foucault, "Mémoire sur le phénomène des interférences entre deux rayons de lumière dans le cas de grandes différences de marche," *ACP*, 26 (1849), 136–148; "Mémoire sur le phénomène des interférences dans le cas de grandes différences de marche, et sur la polarisation chromatique produite par les lames cristallisées épaisses," *ACP*, 30 (1850), 146–159, on 159; Gouy, ref. 34, on 362.

has the form

$$\chi(\omega - \omega_0) = \widetilde{H}(\omega - \omega_0), \tag{26}$$

with $H(t) = 1$ for $t_1 \leq t \leq t_2$ and $H(t) = 0$ for $t < t_1$ or $t > t_2$, t_1 and t_2 being the times that light takes to travel from each extremity of the aperture to the point of observation. The resulting exit amplitude is

$$a(t) \propto \mathrm{e}^{\mathrm{i}\omega_0 t} \left[\int_{t-t_1}^{t-t_2} s(t')\mathrm{e}^{-\mathrm{i}\omega_0 t'} \mathrm{d}t' + \mathrm{e}^{\mathrm{i}\omega_0 \tau} \int_{t+\tau-t_1}^{t+\tau-t_2} s(t')\mathrm{e}^{-\mathrm{i}\omega_0 t'} \mathrm{d}t' \right]. \tag{27}$$

This expression avoids the aforementioned paradox, since it vanishes if the time t is far enough from the period of activity of the source. As Poincaré regarded the oscillating factor $1 + \mathrm{e}^{\mathrm{i}\omega_0 \tau}$ as empirically established by Fizeau and Foucault, he required that

$$\int_{t-t_1}^{t-t_2} s(t')\mathrm{e}^{-\mathrm{i}\omega_0 t'} \mathrm{d}t' = \int_{t+\tau-t_1}^{t+\tau-t_2} s(t')\mathrm{e}^{-\mathrm{i}\omega_0 t'} \mathrm{d}t' \tag{28}$$

for any τ at which the band spectrum is still seen, and he concluded:[36]

> *The experiment of Fizeau and Foucault teaches us... that the luminous motion enjoys a certain kind of permanence expressed in equation* [(28)] ... Thus, a complete analysis leads to exactly the consequences that Mr. Fizeau's clear-sightedness had guessed in advance.

Gouy soon protested: Poincaré had failed to appreciate that the lines of the spectrum could only be separated when the interference delay τ was smaller than the time $t_2 - t_1$ that determines the resolving power of the spectroscope. In this case, the condition (28) is trivially satisfied, no matter how irregular the original motion might be. In sum, the spectroscope is able to produce by itself all the regularity needed to observe interference. Lord Rayleigh and Arthur Schuster had independently come to the same conclusion. This made the nature of white light a matter of speculation. Poincaré silently accepted Gouy's rebuttal. As he later admitted in a different context, "Mathematics are sometimes a hinder, even a danger, when by the precision of their language they induce us to assert more than we know."[37]

---

[36] Poincaré, "Sur le spectre cannelé," *CR*, 120 (1895), 757–762, on 761.

[37] Gouy, "Sur la régularité du mouvement lumineux," *CR*, 120 (1895), 915–917; Rayleigh, "Wave theory of light," in *Encyclopaedia Britannica*, 9th ed. (New York reprint), vol. 24, 421–459, on 425; Arthur Schuster, "On interference phenomena," *Philosophical magazine*, 37 (1894), 509–545; Poincaré, on Pierre Curie and others, *CR*, 143 (1946), 989–998, on 990 [Les mathématiques sont quelques fois une gêne, ou même un danger, quand, par la précision de leur langage, elles nous amènent à affirmer plus que nous ne savons.]. Poincaré's error is the more surprising because in the Hertzian context he had insisted that multiple resonance (in which the resonator plays the role of the spectrometer) was compatible with a damped periodic motion of the electric oscillator.

4. Optics and electromagnetism

Lecturing on Maxwell

Poincaré had planned a course of lectures on the electromagnetic theory of light even before hearing about Hertz's experiments of the winter 1887–88. He delivered this course in the summer of 1888, with a brief mention of Hertz's findings. The fact that Poincaré studied Maxwell's theory and Hertz's contributions in the context of his optical lectures had important implications. On the experimental side, he interpreted Hertz's discovery as a "synthesis of light" and never missed an opportunity to discuss analogies and disanalogies between light and Hertzian waves.[38] On the theoretical side, he paid special attention to the relations between electromagnetism and optics: "I have spent much time studying the relations between electrodynamics and optics," he wrote in 1901 in an analysis of his works.[39] As we will see in a moment, he was at his best when he analyzed the difficulties of conciliating the optics of moving bodies with the electromagnetic theory of light.[40]

Poincaré's comparison between Maxwell's theory and earlier optics caused a conscious turn in his approach to physical theory. Whereas in his optical lectures he described several mechanical models of the ether and their structural interrelations, in his later courses he often relied on general principles such as the energy principle and the principle of least action to guide theoretical construction or to criticize the products of the construction. This new approach, which Poincaré later called the "physics of principles," had roots not only in thermodynamics but also in Maxwell's Treatise on electricity and magnetism of 1873. As was earlier mentioned, in the mature form of this theory, Maxwell avoided any specific ether mechanism and contented himself with requiring the Lagrangian form of the field equations. In the foreword to his lectures on Maxwell's theory, Poincaré described "Maxwell's fundamental idea" as follows:

> *In order to prove the possibility of a mechanical explanation of electricity, we need not worry about finding this explanation itself, we only need to know the expression of the two functions T and U which are the two components of the energy, to form the Lagrange equations for these two*

---

[38]See, e.g., Poincaré, "La lumière et l'électricité d'après Maxwell et Hertz," *Annuaire du Bureau des longitudes* (1894), A1–A22, on A17; *La théorie de Maxwell et les oscillations hertziennes* (Paris, 1899), chap. 11: Imitation des phénomènes optiques, chap. 12: Synthèse de la lumière; Poincaré, ref. 30 (1902), 66.

[39]Poincaré, Analysis of his scientific works, in PO9, 1–14, on 10 [Je me suis beaucoup occupé des rapports entre l'électrodynamique et l'optique].

[40]As is well known, Poincaré actively contributed to the interpretation of Hertz's and related experiments, and he repeatedly lectured on this topic: Poincaré, *Electricité et optique II. Les théories de Helmholtz et les expériences de Hertz* (Sorbonne lectures, 1889–1890), ed. B. Brunhes (Paris, 1891); *Les oscillations électriques* (Sorbonne lectures, 1892–1893), ed. C. Maurain (Paris, 1894). Cf. Buchwald, *The creation of scientific effects: Heinrich Hertz and electric waves* (Chicago, 1994); Michel Atten, *Les théories électriques en France,* 1870–1900. *La contribution des mathématiciens, des physiciens et des ingénieurs à la construction de la théorie de Maxwell.* Thèse de doctorat (Paris: EHESS, 1992).

that matter could not be the sole medium for the propagation of light. Poincaré's *répétiteur* Potier was the man who later corrected Michelson and Morley for miscalculating the path difference in their moving interferometer. In these student years Poincaré performed an ether-drift experiment which he remembered many years later:

> I was long ago a student at the Ecole Polytechnique. I must concede that I am extraordinarily clumsy and that since then I have felt I should better stay away from experimental physics. At that time, however, I was helped by a fellow student, Mr. Favé, who is manually very adroit and who, in addition, has a very resourceful mind. We jointly tried whether the translatory motion of the earth affected the laws of double refraction. If our investigation had led to a positive result, that is, if our light fringes had been shifted, this would only have shown that we lacked experimental skills and that the build up of our apparatus was defective. In reality the outcome was negative, which proved two things at the same time: that the laws of optics are not affected by the translatory motion, and that we were quite lucky on this matter.

In 1888, Poincaré devoted the last chapter of his optical lectures to stellar aberration and other optics of moving bodies. He introduced the Fresnel drag directly in the discussion of stellar aberration, as the drag value compatible with George Biddel Airy's finding (in 1871) that water-filling did not affect the stellar aberration observed in a reflecting telescope. Poincaré then expounded Fizeau's running-water experiment, and gave a general proof that the earth's motion did not affect optical experiments on earth if the ether was dragged by transparent bodies according to Fresnel's hypothesis. Here is a modernized, infinitesimal version of this proof.[44]

The velocity of light with respect to the ether in a substance of optical index n is c/n, if c denotes the velocity of light. The absolute velocity of the ether across this substance is $\alpha\mathbf{v}$, where α is the dragging coefficient and $\mathbf{v}$ the absolute

---

[44]Poincaré, "L'état actuel et l'avenir de la physique mathématique," *Bulletin des sciences mathématiques*, 28 (1904), 302–324, on 320 [C'est l'Astronomie, en somme, qui a soulevé la question en nous faisant connaître l'aberration de la lumière.] (I use the excellent translation in the *Bulletin of the American Mathematical Society*, 12 (1906), 240–260); "Die neue Mechanik," *Himmel und Erde*, 23 (1910), 97–116, on 104 [Ich war damals Schüler der Ecole Polytechnique. Ich muss ihnen gestehen, dass ich außerordentlich ungeschickt bin, und dass ich seitdem gänzlich auf die Experimentalphysik verzichten zu müssen glaubte. Aber zu jener Zeit sprang mir ein Studiengenosse bei, M. Favé, der manuell sehr geschickt und außerdem ein sehr erfinderischer Kopf ist. Wir verbanden uns also zu Untersuchungen, ob die Gesetze der Doppelbrechung durch die Translation der Erde eine störende Modifikation erfahren. Würden unsere Untersuchungen zu einem positiven Resultat geführt haben, d.h. würden unsere Lichtfransen von ihrer Richtung abgelenkt sein, so würde das nur gezeigt haben, dass wir im Experimentieren keine Erfahrung hatten, und dass die Aufstellung unseres Apparates mangelhaft war. Indessen die Untersuchung verlief negativ, und das bewies zwei Dinge zugleich, nämlich dass die Gesetze der Optik durch die Translation nicht gestört werden, und dass wir bei der Sache viel Glück hatten.]; ref. 2, 379–397; Cornu, ref. 2, 101; Hervé Faye, *Cours d'astronomie, 1e division,* 1873–1874 (autographed course, Paris: Ecole Polytechnique), 170–174. On the lycée physics course, cf. Walter, ref. 1.

velocity of the substance (the absolute velocity being defined with respect to the remote, undisturbed parts of the ether). Therefore, the velocity of light along the element dl of an arbitrary trajectory is $c/n + (\alpha - 1)\mathbf{v} \cdot \mathrm{d}\mathbf{l}/\mathrm{d}s$ with respect to the substance (with $\mathrm{d}s = \|\mathrm{d}\mathbf{l}\|$). To first order in u/c, the time taken by light during this elementary travel is

$$\mathrm{d}t = (n/c)\mathrm{d}s + (n^2/c^2)(1-\alpha)\mathbf{v} \cdot \mathrm{d}\mathbf{l}. \tag{29}$$

The choice $\alpha = 1$ (complete drag) leaves the time $\mathrm{d}t$ and the trajectory of minimum time invariant, as should obviously be the case. Fresnel's choice,

$$\alpha = 1 - 1/n^2 \tag{30}$$

yields

$$\mathrm{d}t = (n/c)\mathrm{d}s + (1/c^2)\mathbf{v} \cdot \mathrm{d}\mathbf{l}. \tag{31}$$

If the earth is set to move at the velocity $\mathbf{u}$, the velocity $\mathbf{v}$ is turned into $\mathbf{v}+\mathbf{u}$ so that the time taken by light to travel between two fixed points of the optical setting differs only by a constant from the time it would take if the earth were not moving. Therefore, to first order in u/c interference phenomena are unchanged, and by Fermat's principle of least time the laws of reflection and refraction are also unchanged. Poincaré concluded: "In one word, optical phenomena can provide evidence only for the relative motion of the luminous source and of the ponderable matter with respect to the observer."[45]

Criticizing Lorentz and others

Poincaré returned to the optics of moving bodies in a criticism of Larmor's and Lorentz's electromagnetic theories published in 1895 in l'*Éclairage électrique*. In the spirit of the physics of principles, Poincaré compared three competing electrodynamic theories by Hertz, Helmholtz–Reiff, and Lorentz under three criteria:[46]

1. The theory should account for Fizeau's partial drag.
2. The theory should be compatible with the principle of conservation of electricity.
3. The theory should be compatible with the principle of equality of action and reaction.

Poincaré first showed that Hertz's electrodynamics bodies, being based on the assumption of a fully dragged ether was incompatible with Fizeau's result. Then he argued (erroneously) that the Helmholtz–Reiff theory violated the conservation of electricity.

[45]Poincaré, ref. 2, 389–391 [En un mot les phénomènes optiques ne peuvent mettre en évidence que les mouvements relatifs par rapport à l'observateur de la source lumineuse et de la matière pondérable.]. As Eleuthère Mascart noted in the 1870s, this justification needs to be modified when double refraction and dispersion are taken into account. Poincaré also gave Boussinesq's theory of the Fresnel drag.

[46]Poincaré, "A propos de la théorie de M. Larmor," *L'éclairage électrique*, 3 (1895), 5–13, 289–295; 5 (1895), 5–14, 385–392; also in *PO*9, 369–426, on 395.

Lastly he showed that Lorentz's theory violated the equality of action and reaction, simply remarking that in this theory electromagnetic waves from a remote source could move a charged particle without compensating recoil of the source or of the medium. In Poincaré's opinion, Lorentz's ether, being essentially immobile and being divorced from matter, was too immaterial to carry any momentum:[47]

> I find it difficult to admit that the principle of reaction is violated, even seemingly, and that this principle no longer holds if one considers the actions on ponderable matter and if the reaction of this matter on the ether is left aside.

As no known theory met Poincaré's three criteria and as the one he judged "the least defective," Lorentz's, violated the principle of reaction, Poincaré drew a radical conclusion: "Some day we will have to break the frame into which we endeavor to fit both the optical and the electrical phemomena."[48]

Poincaré further noted that a multitude of optical experiments led to the following "law":

> It is impossible to detect the absolute motion of matter, better said: the relative motion of ponderable matter with respect to the ether. All we can detect is the motion of ponderable matter with respect to ponderable matter.

Lorentz's theory accounted for this impossibility, but only to first order in u/c . As Poincaré knew, "a recent experiment by Michelson" confirmed the law to second order. Poincaré suspected a deep connection between this weakness of Lorentz's theory and its failure to comply with the reaction principle:[49]

> The impossibility of detecting the relative motion of matter with respect to the ether, and the probable equality of action and reaction without taking the action of matter on the ether into account, are two facts that

[47]Ibid., 412 [Il me paraît bien difficile d'admettre que le principe de réaction soit violé, même en apparence, et qu'il ne soit plus vrai si l'on envisage seulement les actions subies par la matière pondérable et si on laisse de côté la réaction de cette matière sur l'éther.]. Poincaré had a similar objection to Larmor's ether, whose momentum density represented the magnetic field. To Oliver Lodge's failure to detect the ether motion caused by an intense magnetic field in Larmor's theory, Poincaré commented (ibid., 382): "Si le résultat avait été positif, on aurait pu mesurer la densité de l'éther et, si le lecteur veut bien me pardonner la vulgarité de cette expression, il me répugne de penser que l'éther soit si arrivé que cela." Cf. Darrigol, "Henri Poincaré's criticism of *fin de siècle* electrodynamics," *Studies in the history and philosophy of modern physics*, 26 (1995), 1–44.

[48]Poincaré, ref. 46, *PO*9, 409 [Il faudra donc un jour ou l'autre briser le cadre où nous cherchons à faire rentrer à la fois les phénomènes optiques et les phénomènes électriques.].

[49]Ibid., 412–413 [Il est impossible de rendre manifeste le mouvement absolu de la matière, ou mieux le mouvement relatif de la matière pondérable par rapport à l'éther; tout ce qu'on peut mettre en évidence, c'est le mouvement de la matière pondérable par rapport à la matière pondérable.][L'impossibilité de mettre en évidence un mouvement relatif de la matière par rapport à l'éther ; et l'égalité qui a sans doute lieu entre l'action et la réaction sans tenir compte de l'action de la matière sur l'éther, sont deux faits dont la connexité semble évidente. Peut-être les deux lacunes seront-elles comblées en même temps.].

seem obviously connected. Perhaps the two defects will be mended at the same time.

Poincaré resumed his Sorbonne lectures on electricity and optics in 1899, with an additional chapter on the optics of moving bodies. There he proved that Lorentz's theory implied the invariance of optical phenomena to first order in u/c except for the time shift $t' = t - ux'/c^2$, which he judged too small to be experimentally observable.[50] Then he described the Michelson–Lorentz experiment and the Lorentz–FitzGerald contraction with the comment:

> This strange property would seem a proper nudge [*coup de pouce*] given by nature in order to prevent that the absolute motion of the earth be revealed by optical phenomena. I cannot be satisfied with this state of affairs. Let me tell you how I see things: I regard it as very probable that optical phenomena depend on the relative motion of the material bodies in presence, optical sources and optical apparatus, and this *not only to second order in the aberration* [*in u/c*] *but rigorously.* As the precision of experiments grows, this principle will be verified in a more precise manner.

With this bet on the exact validity of the relativity principle in optics, Poincaré broke the contemporary consensus that motion through the ether remained in principle detectable at higher orders in u/c. In most physicists' mind, the ether had enough similarity with an ordinary body or substance to disturb optical phenomena when the earth rushed through it. In Poincaré's mind, the ether was too ethereal to be physically detectable. Poincaré therefore criticized Lorentz's order-by-order compensations as provisional subterfuges to be replaced by a better theory:[51]

> Shall we need a new *coup de pouce*, a new hypothesis, at each order of approximation? Evidently no: a well-wrought theory should allow us to demonstrate the principle [that optical phenomena depend only on the relative motion of the implied material bodies] in one stroke and in full rigor. Lorentz's theory does not do that yet. Of all existing theories, it

[50] Poincaré was also aware of the "Liénard force," which is a first-order correction to the Lorentz force in a moving system.

[51] Poincaré, *Electricité et optique. La lumière et les théories électrodynamiques* (Sorbonne lectures of 1888, 1890, and 1899, plus the text of Poincaré, ref. 46), ed. by J. Blondin and E. Néculcéa (Paris, 1901), 536 [Cette étrange propriété semblerait un véritable "*coup de pouce*" donné par la nature pour éviter que le mouvement absolu de la terre puisse être révélé par les phénomènes optiques. Cela ne saurait me satisfaire et je crois devoir dire ici mon sentiment : je regarde comme très probable que les phénomènes optiques ne dépendent que des mouvements relatifs des corps matériels en présence, sources lumineuses ou appareils optiques et *cela non pas aux quantités près de l'ordre du carré ou du cube de l'aberration mais rigoureusement.* A mesure que les expériences deviendront plus exactes, ce principe sera vérifié avec plus de précision.] (Poincaré's emphasis) [Faudra-t-il un nouveau coup de pouce, une hypothèse nouvelle, à chaque approximation? Evidemment non: une théorie bien faite devrait permettre de démontrer le principe d'un seul coup dans toute sa rigueur. La théorie de Lorentz ne le fait pas encore. De toutes celles qui ont été proposées, c'est elle qui est le plus près de le faire. On peut donc espérer de la rendre parfaitement satisfaisante sous ce rapport sans la modifier trop profondément.].

> is the theory that is closest to this aim. We may therefore hope to make it completely satisfactory in this regard without altering it too much.

Poincaré expressed similar views in his address to the international congress of physics in Paris in 1900. After describing the fate of the imponderable fluids of older physics, he asked: "And our ether, does it really exist?" He listed a few circumstances in favor of the ether: the elimination of direct action at a distance; Fizeau's experiment, which seemed to require the interplay of two different media, the ether and the running water ("We seem to be fingering the ether"); the violation of the principle of reaction in Lorentz's theory, which seemed to require the ether to carry momentum; and the predicted effects of the ether wind in this theory. Poincaré next recalled that experimenters had failed to detect the latter effects, and he argued that this failure could not be accidental:[52]

> Experiments were performed in which the first-order terms should have been detected. The results were negative. Was it by chance? No one believed so. A general explanation was sought for, and Lorentz found it. He showed that the first-order terms cancelled each other; but this was not true for the second-order terms. Then more precise experiments were done. They were also negative. Again, that could not be by chance. An explanation was needed. It was found. Explanations can always be found: Of hypotheses there is never a lack.
> This is not enough. Who would not see that chance still plays a too large part? Is it not a singular coincidence if a certain circumstance comes forth to destroy the first-order terms and then another, completely different but equally opportune circumstance takes care of the second-order terms? No, we must find a common explanation for both kinds of terms; and then there is ample reason to believe that this explanation will also work for terms of higher order and that the compensation will be rigorous and absolute.

[52]Poincaré, "Sur les rapports de la physique expérimentale et de la physique mathématique," in C.E. Guillaume and L. Poincaré (eds.), *Rapports présentés au congrès international de physique* (Paris, 1900), vol. 4, 1–29, on 21–22 [Et notre éther, existe-t-il vraiment?] [On a fait des expériences qui auraient dû déceler les termes du premier ordre; les résultats on été négatifs; cela pouvait-il être par hasard? Personne ne l'a admis; on a cherché une explication générale, et Lorentz l'a trouvée; il a montré que les termes du premier ordre devaient se détruire, mais il n'en était pas de même de ceux du second. Alors on a fait des expériences plus précises; elles ont aussi été négatives; ce ne pouvait non plus être l'effet du hasard ; il fallait une explication ; on l'a trouvée ; on en trouve toujours ; les hypothèses, c'est le fonds qui manque le moins. – Mais ce n'est pas assez; qui ne sent que c'est encore là laisser au hasard un trop grand rôle? Ne serait-ce pas aussi un hasard que ce singulier concours qui ferait qu'une certaine circonstance viendrait juste à point pour détruire les termes du premier ordre, et qu'une autre circonstance, tout à fait différente, mais tout aussi opportune, se chargerait de détruire ceux du second ordre? Non, il faut trouver une même explication pour les uns et pour les autres, et alors tout nous porte à penser que cette explication vaudra également pour les termes d'ordre supérieur, et que la destruction mutuelle de ces termes sera rigoureuse et absolue.]

Analyzing the crisis

In the same year 1900, Poincaré contributed a memoir on "Lorentz's theory and the principle of reaction" to a volume celebrating Lorentz's jubilee. There he developed his earlier idea that Lorentz's theory implied an intolerable violation of the reaction principle. The Maxwell–Lorentz equations for a system of charged particles of mass m moving at the velocity $\mathbf{v}$ and interacting through the fields $\mathbf{E}$ and $\mathbf{B}$ leads to the equation

$$\sum m\mathbf{v} + \int c^{-1}\mathbf{E} \times \mathbf{B} d\tau = \text{constant}. \tag{32}$$

Consequently, the momentum of matter is not conserved. Unlike British field theorists, Poincaré refused to interpret the integral of $\mathbf{E} \times \mathbf{B}/c$ as the ether's momentum, even though he showed that the theorem of the center of mass could be saved by regarding this vector as the momentum density of a fictitious fluid moving at the velocity c and created or annihilated by the sources. He indeed believed that any violation of the principle of reaction *when applied to matter alone* led to absurdities. By an argument borrowed from Newton's *Principia*, if the action of a material body A on a material body B differs from the reciprocal action, a rigid combination of A and B would forever accelerate and perpetual motion would thus become possible. Poincaré noted, this argument presupposes that the net force on the combined body is independent of the acquired motion, in conformity with the relativity principle. In general, Poincaré believed that any violation of the reaction principle had to be intimately related to a violation of the relativity principle. In order to confirm this interconnection in the case of electromagnetic interactions, he considered a Hertzian oscillator placed at the focus of a parabolic mirror and emitting radiation at a constant rate. This system moves with the absolute velocity $\mathbf{u}$ in the direction of emission, and is heavy enough so that the change of this velocity can be neglected during its recoil. For an observer at rest in the ether, the conservation of energy reads

$$S = J + (-J/c)u, \tag{33}$$

where S is the energy spent by the oscillator in a unit time, J the energy of the emitted wave train, and $-J/c$ the recoil momentum according to Lorentz's theory. For an observer moving at the velocity $\mathbf{u}$ of the emitter, the recoil force does not work, and the spent energy S is obviously the same. According to the Lorentz transformations for time and fields (to first order), this observer should ascribe the energy $J(1 - u/c)$ to the emitted radiation and the value $(-J/c)(1 - u/c)$ to the recoil momentum. Hence the energy principle is satisfied for the moving observer, but the (time integral of the) electromagnetic force acting on the emitter is modified by Ju/c^2. Poincaré regarded this difference as a first-order violation of the relativity principle, the expected counterpart of the first-order violation of the principle of reaction.[53]

---

[53]Poincaré, "La théorie de Lorentz et le principe de la réaction," in *Recueil de travaux offerts par les auteurs à H.A. Lorentz à l'occasion du 25ème anniversaire de son doctorat le* 11 *décembre*

In this calculation, Poincaré used Lorentz's first-order field transformations, including the local time $t' = t - ux'/c^2$, which he defined in the following manner:[54]

> I suppose that observers placed in different points [of the moving frame] set their watches by means of optical signals; that they try to correct these signals by the transmission time, but that, ignoring their translational motion and thus believing that the signals travel at the same speed in both directions, they content themselves with crossing the observations, by sending one signal from A to B, then another from B to A. The local time is the time given by watches adjusted in this manner.

Poincaré only made this remark *en passant*, gave no proof, and did as if it belonged to Lorentz. The proof goes as follows. When B receives the signal from A, he sets his watch to zero (for example), and immediately sends back a signal to A. When A receives the latter signal, he notes the time τ that has elapsed since he sent his own signal, and sets his watch to the time $\tau/2$. By doing so he commits an error $\tau/2 - t_-$, where t_- is the time that light really takes to travel from B to A. This time, and that of the reciprocal travel are given by

$$t_- = \mathrm{AB}/(c+u) \text{ and } t_+ = \mathrm{AB}/(c-u), \qquad (34)$$

since the velocity of light is c with respect to the ether (see Fig. 4). The time τ is the sum of these two traveling times. Therefore, to first order in u/c the error committed in setting the watch A is

$$\tau/2 - t_- = (t_+ - t_-)/2 = u\mathrm{AB}/c^2. \qquad (35)$$

At a given instant of the true time, the times indicated by the two clocks differ by $u\mathrm{AB}/c^2$, in conformity with Lorentz's expression of the local time.

Poincaré transposed this synchronization procedure from an earlier discussion on the measurement of time, published in 1898. There he noted that the dating of astronomical events was based on the implicit postulate "that light has a constant velocity, and in particular that its velocity is the same in all directions." He also explained the optical synchronization of clocks at rest, and mentioned its similarity with the telegraphic synchronization that was then being developed for the purpose of longitude measurement. In his interpretation of Lorentz's local time, Poincaré simply transposed this procedure to moving clocks and to mov-

---

1900, *Archives néerlandaises*, 5 (1900), 252–278. In 1898, Alfred Liénard had already noted the first-order modification of the Lorentz force through a Lorentz transformation. In relativity theory, this modification is compensated by the variation of the mass of the emitter. Cf. Darrigol, "Poincaré, Einstein, et l'inertie de l'énergie," *CR*, 1 (2000), 143–153; "The genesis of the theory of relativity," in T. Damour, O. Darrigol, B. Duplantier, V. Rivasseau (eds.), *Einstein* 1905–2005*: Poincaré seminar* 2005 (Basel : Birkhäuser, 2006), 1–31.

[54]Poincaré, ref. 53, on 272 [Je suppose que des observateurs placés en différents points, règlent leurs montres à l'aide de signaux lumineux ; qu'ils cherchent à corriger ces signaux du temps de la transmission, mais qu'ignorant le mouvement de translation dont ils sont animés et croyant par conséquent que les signaux se transmettent également vite dans les deux sens, ils se bornent à croiser les observations, en envoyant un signal de A en B, puis un autre de B en A. Le temps local t' est le temps marqué par les montres ainsi réglées.]

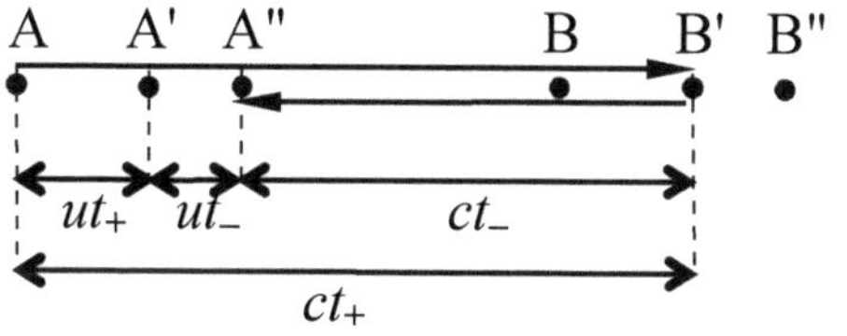

FIGURE 4. Cross-signaling between two observers moving at the velocity u through the ether. The points A, A', A'', B, B', B'' represent the successive positions of the observers in the ether when the first observer sends a light signal, when the second observer receives this signal and sends back another signal, and when the first observer receives the latter signal.

ing observers who could not know their motion through the ether and therefore could only do as if the velocity of light was a constant in their own frame. As we will see, the metrological function of light had a growing importance in Poincaré's understanding of relativity theory.[55]

Poincaré generalized his worries about violations of general principles in 1904, in a talk delivered during the international exhibition in Saint-Louis, Missouri. There he described the transition from a physics of central forces and mechanical models to a physics of general principles mainly based on the energy principle, Carnot's principle, the relativity principle (so named for the first time), the principle of reaction, Lavoisier's principle of the conservation of mass, and the principle of least action. Poincaré proceeded to show that all these principles, except the last, were in danger. On the relativity principle, he had to say:

> This principle is not only confirmed by our daily experience, not only is it the necessary consequence of the hypothesis of central forces, but it appeals to our common sense with irresistible force. And yet it also is being fiercely attacked.

As Poincaré explained, the threat did not come from experiments but from theory, because even the most successful theory of electromagnetism and optics, Lorentz's theory, could only save the principle approximately and at the price of more or less artificial assumptions: "Thus the principle of relativity has in recent times been valiantly defended; but the very vigor of the defense shows how serious was the attack." Among Lorentz's relativity-saving assumptions, Poincaré singled out the "most ingenious" one, the local time, which he again justified by optical synchronization.[56]

[55] Poincaré, "La mesure du temps," *Revue de métaphysique et de morale*, 6 (1898), 371–384. On the telegraphic context, cf. Peter Galison, *Einstein's clocks, Poincaré's maps: Empires of time* (New York, 2003).

[56] Poincaré, ref. 44 (1904), 310 [Celui-là non seulement est confirmé par l'expérience quotidienne, non seulement il est une conséquence nécessaire de l'hypothèse des forces centrales, mais il s'impose à notre bon sens d'une façon irrésistible; et pourtant lui aussi est battu en brèche.],

Poincaré next considered the violation of the principle of reaction in Lorentz's theory and the ad hoc character of attempts to save it by making the ether a momentum carrier:

> We might also suppose that the motions of ordinary matter were exactly compensated by motions of the ether... The principle, if thus interpreted, could explain anything since whatever the visible motions we could imagine hypothetical motions to compensate them. But if it can explain anything, it will allow us to foretell nothing; it will not allow us to choose between the various possible hypotheses, since it explains everything in advance. It therefore becomes useless.

Here we have a glimpse at Poincaré's refined conception of physical principles: they originate in the generalization of a multitude of experimental results; then, owing to this generality, they tend to be regarded as conventions and thus become immune to refutation; however, they can be endangered when the strategies used to save them become too artificial. In Poincaré's view, the momentum-carrying ether belonged to this sort of degenerative strategy. "This is why," he went on, "I have for a long time thought that the consequences of the theories that contradict Newton's principle [of reaction] would some day be abandoned; and yet the recent experiments on the motion of the electrons emitted by radium seem rather to confirm them."[57]

Poincaré was here alluding to Walther Kaufmann's experiments on the electric and magnetic deflection of fast electrons from radium, and to Max Abraham's explanation of these results in a new dynamics of the electron in which the mass of the electron resulted from the inertia of the accompanying electromagnetic field and therefore depended on the electron's velocity. Interestingly, Abraham obtained this theory by reinterpreting Poincaré's fictitious-fluid momentum ($\mathbf{E} \times \mathbf{B}/c$) of 1900 as a true field momentum and by integrating this momentum for the self-field of the electron. Despite his original commitment to the reaction principle, Poincaré

---

312 [Ainsi le principe de la relativité a été dans ces derniers temps vaillamment défendu, mais l'énergie même de la défense prouve combien l'attaque était sérieuse.]

[57]Ibid., 314 [On peut supposer aussi que les mouvements de la matière proprement dite sont exactement compensés par ceux de l'éther, mais cela nous amènerait aux mêmes réflexions que tout à l'heure. Le principe ainsi entendu pourra tout expliquer, puisque, quels que soient les mouvements visibles, on aura toujours la faculté d'imaginer des mouvements hypothétiques qui les compensent. Mais, s'il peut tout expliquer, c'est qu'il ne nous permet de rien prévoir, il ne nous permet pas de choisir entre les différentes hypothèses possibles, puisqu'il explique tout d'avance. Il devient donc inutile.] [C'est pourquoi j'ai longtemps pensé que ces conséquences de la théorie, contraires au principe de Newton, finiraient un jour par être abandonnées et pourtant les expériences récentes sur les mouvements des électrons issus du radium semblent plutôt les confirmer.]

was willing to bow to the new experimental evidence. He even sketched a future relativistic mechanics:[58]

> From all these results, if they were to be confirmed, would issue a wholly new mechanics which would be characterized above all by the fact that there could be no velocity greater than that of light, any more than a temperature below that of absolute zero. For an observer participating in a motion of translation of which he has no suspicion, no apparent velocity could surpass that of light, and this would be a contradiction, unless one recalls the fact that this observer does not use the same sort of timepiece as that used by a stationary observer, but rather a watch giving the 'local time.'

Poincaré formulated most of his arguments in the conditional mode, not being sure that the threatening experimental results were definitive or that the endangered theory had lost all its steam. His advice to fellow theorists was moderate: "All hope of obtaining better results is not yet lost. Let us, then, take the theory of Lorentz. Let us turn it over and over, let us modify it little by little, and all will be well, perhaps." Poincaré offered one concrete suggestion:

> Instead of supposing that bodies in motion undergo a contraction in the direction of motion and that this contraction is the same whatever the nature of these bodies and the forces to which they are subjected, could not a simpler and more natural hypothesis be made? One might suppose, for example, that it is the ether which changes when it is in relative motion with respect to the material substance which permeates it; that, thus modified, it no longer transmits the disturbances with the same velocity in all directions. It would transmit more rapidly those disturbances which are being propagated parallel to the motion of the substance, be it in the same direction or in the opposite, and less rapidly those which are propagated at right angles. The wave surfaces would then no longer be spheres, but ellipsoids, and one could do without this extraordinary contraction of all bodies. I am giving this only by way of example, since the modifications that could be tried are evidently susceptible of infinite variation.

In this speculation, Poincaré seems to be taking the ether more seriously than he had done earlier, in conformity with his new willingness to make the ether a momentum carrier. We will return to this point in a moment.[59]

---

[58]Ibid., 316–317 [De tous ces résultats, s'ils se confirmaient, sortirait une mécanique entièrement nouvelle qui serait surtout caractérisée par ce fait qu'aucune vitesse ne pourrait dépasser celle de la lumière... Pour un observateur, entraîné lui-même dans une translation dont il ne se doute pas, aucune vitesse apparente ne pourrait non plus dépasser celle de la lumière; et ce serait là une contradiction, si l'on ne se rappelait que cet observateur ne se servirait pas des mêmes horloges qu'un observateur fixe, mais bien d'horloges marquant le "temps local."]

[59]Ibid., 319–320 [Tout espoir d'obtenir de meilleurs résultats n'est pas encore perdu. Prenons donc la théorie de Lorentz, retournons-la dans tous les sens: modifions-la peu à peu, et tout s'arrangera peut-être.] [Ainsi, au lieu de supposer que les corps en mouvement subissent une

The dynamics of the electron

In the same year 1904, Lorentz perfected his theory in a way that answered at least one of Poincaré's objections: Lorentz now obtained the invariance of optical phenomena at every order in u/c (though only in the dipolar approximation) and through a single transformation of his equations, granted that all forces (including cohesion forces) behaved like electromagnetic forces and that electrons underwent the Lorentz contraction during their motion through the ether. As he had earlier done to first and second order, Lorentz first applied a Galilean transformation to the Maxwell–Lorentz equations in order to take into account the motion of the earth through the ether, and then applied a second transformation, which, combined with the first, brought back the equations to (nearly) the same form as they had in the ether frame. The combined transformation is nearly what Poincaré later called a Lorentz transformation. Lorentz called the transformed field states "corresponding states" and used them only as a formal intermediate step toward the true physical states in the ether. He had not caught Poincaré's hint that the transformed states were those perceived by moving observers under natural conventions.[60]

With the eye of a group theorist, Poincaré immediately saw that with a few minor corrections Lorentz's transformations exactly preserved the form of the Maxwell–Lorentz equations. From this formal invariance, he concluded that under Lorentz's assumptions all optical and electrodynamical phenomena complied with the relativity principle:

> Lorentz's idea may be summarized thus: The reason why a common translatory motion can be imparted to the entire system without any alteration of the observable phenomena, is that the equations of an electromagnetic medium are unaltered by certain transformations which we shall call Lorentz transformations. In this way two systems, of which one is fixed and the other is in translatory motion, become exact images of each other.

Poincaré developed and exploited the Lie-group structure of the transformations in his subsequent analysis of the various models of the electron. In an action-based formulation, he determined that Lorentz's model of the contractile electron,

---

contraction dans le sens du mouvement et que cette contraction est la même quelles que soient la nature de ces corps et les forces auxquelles ils sont d'ailleurs soumis, ne pourrait-on pas faire une hypothèse plus simple et plus naturelle? On pourrait imaginer, par exemple, que c'est l'éther qui se modifie quand il se trouve en mouvement relatif par rapport au milieu matériel qui le pénètre, que, quand il est ainsi modifié il ne transmet plus les perturbations avec la même vitesse dans tous les sens. Il transmettrait plus rapidement celles qui se propageraient parallèlement au mouvement au milieu, soit dans le même sens, soit en sens contraire, et moins rapidement celles qui se propageraient perpendiculairement. Les surfaces d'onde ne seraient plus des sphères, mais des ellipsoïdes et l'on pourrait se passer de cette extraordinaire contraction de tous les corps. Je ne cite cela qu'à titre d'exemple, car les modifications que l'on pourrait essayer seraient évidemment susceptibles de varier à l'infini.]

[60] Hendrik Antoon Lorentz, "Electromagnetic phenomena in a system moving with any velocity smaller than light," Royal Academy of Amsterdam, *Proceedings* (1904), 809–831.

completed with proper cohesive stresses, was the only one compatible with the Lorentz-group symmetry. Lastly, he sketched a Lorentz covariant theory of gravitation in which the gravitational force propagated with the velocity of light.[61]

In the spring of 1905, Poincaré modestly announced his *Dynamique de l'électron* as a gloss on Lorentz's results: "I have be led to modify and complete them only with regard to a few details." Yet much of his theory was novel and important: the exact invariance, its relativistic interpretation, its group-theoretical formulation, and its application to non-electromagnetic forces of cohesion or gravitation. The group-theoretical aspects are especially impressive, for they inaugurated a now pervasive style of theoretical physics. On the interpretive side, Poincaré's introduction to his memoir shows that he was not as close to modern relativity as his formal considerations would suggest. After discussing the plausibility of his covariant theory of gravitational forces, he remarked:[62]

> Even assuming, however, that [the new theory turns out to agree with astronomical tests], what conclusion should we draw? If the attraction is propagated with the velocity of light, this cannot be by a fortuitous occurrence; it must be the expression of a function of the ether. Then we will have to investigate the nature of this function and to relate it to the other functions of the fluid [ether]. We cannot be satisfied with formulae that are merely placed side by side and agree only by a lucky chance. These formulae must, as it were, interlock. The mind will consent only

[61] "Sur la dynamique de l'électron," *Rendiconti del Circolo Matematico di Palermo*, 21 (1906), 129–176, on 130 [L'idée de LORENTZ peut se résumer ainsi: si on peut, sans qu'aucun des phénomènes apparents soit modifié, imprimer à tout le système une translation commune, c'est que les équations d'un milieu électromagnétique ne sont pas altérées par certaines transformations, que nous appellerons *transformations de* LORENTZ; deux systèmes, l'un immobile, l'autre en translation, deviennent ainsi l'image exacte l'un de l'autre.]. Cf. Jean-Pierre Provost and Christain Bracco, "La théorie de la relativité de Poincaré de 1905 et les transformations actives," *Archive for the history of exact sciences*, 60 (2006), 337–351; "De l'électromagnétisme à la mécanique : le rôle de l'action dans le mémoire de Poincaré de 1905," *Revue d'histoire des sciences*, 62 (2009), 457–493; Michel Le Bellac, "The Poincaré group," in E. Charpentier, E. Ghys, and A. Lesne (eds.), *The scientific legacy of Poincaré* (London, 2010), 329–350. On Poincaré's attempt at a relativistic theory of gravitation, cf. Walter, "Breaking in the 4-vectors: The four-dimensional movement in gravitation, 1905–1910," in Jürgen Renn and Matthias Schemmel (eds.), *The genesis of general relativity*, 4 vols. (Berlin, 2007), vol. 3, *Gravitation in the twilight of classical physics: Between mechanics, field Theory, and astronomy*, 193–252.

[62] Poincaré, "Sur la dynamique de l'électron," *CR*, 140 (1905), 1504–1508, on 1505 [J'ai été seulement conduit à les modifier et à les compléter sur quelques points de détail.]; ref. 61, 131 [Mais en admettant même que cette discussion tourne à l'avantage de la nouvelle hypothèse, que devrons-nous conclure? Si la propagation de l'attraction se fait avec la vitesse de la lumière, cela ne peut être par une rencontre fortuite, cela doit être parce que c'est une fonction de l'éther; et alors il faudra chercher à pénétrer la nature de cette fonction, et la rattacher aux autres fonctions du fluide. Nous ne pouvons nous contenter de formules simplement juxtaposées et qui ne s'accorderaient que par un hasard heureux; il faut que ces formules arrivent pour ainsi dire à se pénétrer mutuellement. L'esprit ne sera satisfait que quand il croira apercevoir la raison de cet accord, au point d'avoir l'illusion qu'il aurait pu le prévoir.]

> when it sees the reason for the agreement, to the point of fancying that it could have predicted this agreement.

Poincaré went on to compare his and Lorentz's theory to the Ptolemaic stage of cosmology. He speculated that a revision of light-based metrology might bring a sort of Copernican revolution in the contemporary theories of matter, electricity, and gravitation:

> Now, there may be an analogy with our problem. If we assume the relativity postulate, we find a number common to the law of gravitation and the laws of electromagnetism, and this number is the velocity of light; and this same number should appear in every other force, of whatever origin. There can be only two explanations for this state of affairs:
> - Either everything in the universe is of electromagnetic origin.
> - Or this constituent which is common to all the phenomena of physics is only an appearance [*une apparence*], something that comes from our methods of measurement. How do we measure? By the congruence of objects regarded as rigid, one might first reply. But this is no longer so in our present theory, if the Lorentz contraction is assumed. In this theory, two lengths are by definition equal if they are traversed by light in the same time. Perhaps by abandoning this definition Lorentz's theory would be as deeply changed as Ptolemy's system was by Copernicus' intervention.

The prominence of Poincaré's optical concerns is evident in his extract: in his relativity theory, the velocity of light is the unifying parameter. This fact may be explained in two manners: either by making the optical (electromagnetic) ether the bearer of every interaction as was done in the electromagnetic worldview of some of Poincaré's contemporaries, or by renouncing the light-based metrology that leads to the contraction of lengths through the null result of the Michelson–Morley experiment. The second alternative, the one Poincaré compares to the Copernican revolution, has sometimes been interpreted as an anticipation of Einstein's version of the relativity theory. More likely, Poincaré had in mind something like the heterogeneous ether suggested in his Saint-Louis lecture.[63]

---

[63] Ibid., 131–132 [Ici il est possible qu'il y ait quelque chose d'analogue; si nous admettions le postulat de relativité, nous trouverions dans la loi de gravitation et dans les lois électromagnétiques un nombre commun qui serait la vitesse de la lumière; et nous le retrouverions encore dans toutes les autres forces d'origine quelconque, ce qui ne pourrait s'expliquer que de deux manières: – Ou bien il n'y aurait rien au monde qui ne fût d'origine électromagnétique. – Ou bien cette partie qui serait pour ainsi dire commune à tous les phénomènes physiques ne serait qu'une apparence, quelque chose qui tiendrait à nos méthodes de mesure. Comment faisons-nous nos mesures? En transportant, les uns sur les autres, des objets regardés comme des solides invariables, répondra-t-on d'abord; mais cela n'est plus vrai dans la théorie actuelle, si l'on admet la contraction lorentzienne. Dans cette théorie, deux longueurs égales, ce sont, par définition, deux longueurs que la lumière met le même temps à parcourir. – Peut-être suffirait-il de renoncer à cette définition, pour que la théorie de Lorentz soit aussi complètement bouleversée que l'a été le système de Ptolémée par l'intervention de Copernic.].

Light and measure

Poincaré's memoir on the dynamics of the electron was a highly mathematical, unusually long and difficult work published in a mathematical journal, the *Rendiconti del Circolo Matematico di Palermo.* In this text Poincaré did not address the problem of measurement in relativity theory, save for the speculation just mentioned and for a brief mention of stellar aberration and the Michelson–Morley experiment. He did not even repeat his earlier argument that Lorentz's local time represented the time measured by moving observers. There may be several reasons for this silence: Poincaré did not have yet the generalization of his first-order argument to higher orders; his memoir was focused on the dynamics of the electron, not on the effects of the earth motion; it did not involve any change of reference frame, as Poincaré interpreted the Lorentz-transformed field states as the states of a globally boosted physical system with respect to the ether frame; in this active view of the Lorentz transformation, one may imagine that the space- and time-measuring agencies belong to the boosted system, in which case the invariance of the equations describing the global system implies that the transformed field and coordinates are the measured ones.

Poincaré first addressed the metrological aspects of relativity theory in his Sorbonne lectures of winter 1906–07 on "The limits of Newton's law." He did this in the eleventh chapter on the dynamics of electron, beginning with a discussion of stellar aberration. In conformity with his long familiarity with this phenomenon, Poincaré had introduced both the *Comptes rendus* summary and the Palermo memoir on the dynamics of the electron with the assertion that aberration seemed to allow for the determination of the velocity of the earth through the ether, not the relative velocity of the earth with respect to the observed star.[64] The explanation, first given in his Sorbonne lectures of 1906, runs as follows. On Fig. 5, $\overrightarrow{OA}$ and $\overrightarrow{OA_1}$, with $OA = OA_1 = c$, represent the absolute velocity of light for two stars diametrically opposed on the celestial sphere ("absolute" here means "with respect to the ether"); $\overrightarrow{AB} = \overrightarrow{A_1B_1}$ represents the opposite of the absolute velocity of the sun; $\overrightarrow{BC} = \overrightarrow{B_1C_1}$ the opposite of the velocity of the earth with respect to the sun at a given time of the year; $\overrightarrow{BC'} = \overrightarrow{B_1C'_1}$ the same velocity six months later (all vectors are in the same plane). The first star is observed in the directions of the vectors $\overrightarrow{OC}$ and $\overrightarrow{OC'}$, the second in the directions of the vectors $\overrightarrow{OC_1}$ and $\overrightarrow{OC'_1}$. As the angles COC' and $C_1OC'_1$ are generally different, the angular amplitude of the apparent oscillation of the two stars in the sky are different; this difference is a second-order effect of the absolute velocity of the sun.[65]

---

[64]The first occurrence of this assertion is in the Saint-Louis lecture, ref. 44 (1904), 320.

[65]Poincaré, "Les limites de la loi de Newton," Sorbonne lectures of Winter 1906–1907, ed. by M. Chopinet after notes taken by Henri Vergne, in *Bulletin astronomique publié par l'observatoire de Paris*, 17 (1953), 121–365, on 216–217 (Vergne's original notes have recently been found at the Bordeaux Observatory; I thank Scott Walter for showing me the pages concerning the light ellipsoid; these do not significantly differ from the printed version); "La dynamique de l'électron," *Revue générale des sciences pures et appliquées*, 19 (1908), 386–402, on 390–391; "La mécanique

Dropping the + index and calling x the abscissa of the second observer, we also have

$$x' = \gamma(x - ut), \tag{42}$$

which, together with the former equation, implies

$$t' = \gamma(t - ux/c^2). \tag{43}$$

Poincaré's ellipsoid can thus be used to derive the Lorentz transformations. Poincaré, however, rather saw the light ellipsoid as a simple geometrical means to prove that the Lorentz contraction implied the apparent isotropy of light propagation for observers using optically synchronized clock and contracted rulers. In turn, this isotropy implied the invariance of optical phenomena.[69]

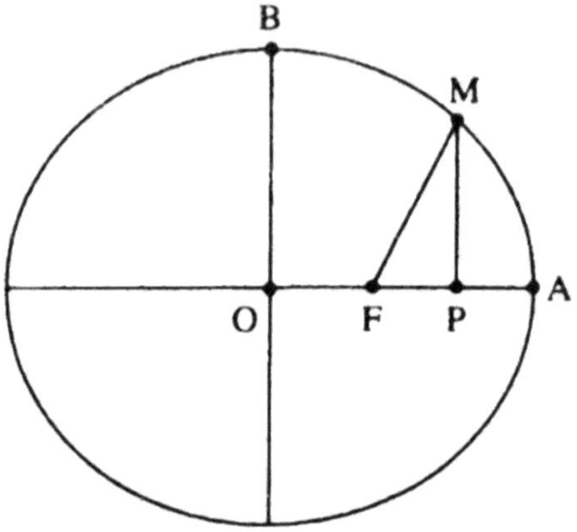

FIGURE 6. Poincaré's light ellipsoid ($a = \mathrm{OA}, b = \mathrm{OB}, f = \mathrm{OF}$).

As Poincaré explained in the following paragraph of his Sorbonne lectures of 1906–07, the Lorentz transformations could be regarded as connecting the field equations in the ether frame to the apparent field equations that relate the apparent field quantities and the apparent space and time coordinates in a moving frame. Apparent quantities are those measured by moving observers ignoring their motion

[69]Poincaré, ref. 65 (1908), 393. The two last equations on that page translate into $\tau = t - \gamma ux'/c^2$ and FM $= \gamma^{-1}ct$. There is an obvious misprint in the latter equation: it should be FM $= \gamma^{-1}c\tau$. Moreover, Poincaré calls τ "the apparent duration of transmission," whereas it is only proportional to the apparent time. This loose terminology probably results from Poincaré's focus on the isotropy of apparent propagation. His commentary to FM $= \gamma^{-1}c\tau$ indeed reads: "Namely, the *apparent* duration of transmission is proportional to the *apparent* distance. This time, the compensation is *exact*, and this is the explanation of Michelson's experiment." Poincaré improved this aspect of his reasoning in lectures delivered in July 1912 at the Ecole Supérieure des Postes et des Télégraphes: *La dynamique de l'électron*, ed. by Viard and Pomey (Paris, 1913), 45–46. In this last occasion (on 44), Poincaré briefly mentioned that "mechanical phenomena were accelerated by a translatory motion." Otherwise, he never discussed time dilation: cf. Thibault Damour, *Si Einstein m'était conté* (Paris, 2005), Chapter 1.

through the ether and therefore misestimating the true quantities defined in the ether frame. The difference with Einstein's theory is obvious: although Poincaré, like Einstein, assumed the complete invariance of *observable phenomena* when passing from an inertial frame to another, he did not require the invariance of the *theoretical description* of the phenomena. In his view, the ether frame was a privileged frame in which true space and time were defined. The Lorentz-transformed quantities in another frame were only "apparent." Of course, the choice of the ether frame could only be conventional, since the relativity principle excluded any empirically detectable difference between the various inertial frames.[70]

This attitude explains some features of Poincaré's light ellipsoid that may seem very odd to a modern, Einsteinian reader. For Poincaré, like for Lorentz, the Lorentz contraction is meant to be a physical effect of the motion of a material body through the ether; it is not the sort of perspectival effect conceived by Einstein; measuring a light pulse at a given value of the true time with contracted rods is a natural operation because the pulse is meant to be a disturbance of the ether and because the contraction is caused by the motion of the rulers through the ether. Although this point of view contradicts Einsteinian intuitions, it is self-consistent and it leads to the correct expression of the Lorentz transformations.

The ether

One may still wonder why Poincaré chose to maintain a rigorously undetectable ether. Had he not written, some twenty years earlier, that "probably the ether will some day be thrown aside as useless"? As a supporter of Felix Klein's definition of geometry through a group of transformation, was not he prepared to define a new geometry based on the Lorentz group? Poincaré answered this question in a talk given toward the end of his life: he was perfectly aware of the possibility (exploited by Hermann Minkowski) of defining the geometry of spacetime through the Lorentz group; at the same time, he still believed that the choice of the group defining a geometry was largely conventional and that ancestral habits were important in judging the convenience of a convention:[71]

> What should be our position with regard to these new conceptions? Shall we be compelled to modify our conclusions? No, assuredly: we

[70]Ibid., 221. For a comparison between Einstein's and Poincaré's approaches, cf. Darrigol, ref. 53; "The mystery of the Einstein Poincaré connection," *Isis*, 95 (2004), 614–626.

[71]Poincaré, "L'espace et le temps," *Scientia*, 12 (1912), 159–170, on 170 [Quelle va être notre position en face de ces nouvelles conceptions ? Allons-nous être forcés de modifier nos conclusions ? Non certes: nous avions adopté une convention parce qu'elle nous semblait commode, et nous disions que rien ne pourrait nous contraindre à l'abandonner. Aujourd'hui certains physiciens veulent adopter une convention nouvelle. Ce n'est pas qu'ils y soient contraints ; ils jugent cette convention nouvelle plus commode, voilà tout ; et ceux qui ne sont pas de cet avis peuvent légitimement conserver l'ancienne pour ne pas troubler leurs vieilles habitudes. Je crois, entre nous, que c'est ce qu'ils feront encore longtemps.]. On Minkowski's views, see Walter, "Minkowski, mathematicians, and the mathematical theory of relativity," in H. Goenner, J. Renn, J. Ritter, and T. Sauer (eds.), *The expanding Worlds of General Relativity* (*Einstein Studies* 7) (Boston, 1999), 45–86.

> had adopted a convention because it seemed convenient to us and we were saying that nothing could force us to abandon it. Today some physicists want to adopt a new convention. Not that they are forced to do so. They simply judge this convention to be more convenient. Those who do not share this opinion can legitimately keep the old convention in order not to disturb their old habits. Between you and mę, I believe they will do so for a long time.

For those who wanted to preserve the old conventions for space and time, the ether was still useful as a reference frame for true space and time. There is an additional reason for Poincaré's preservation of the ether. Until 1900, the only role he gave to the ether was the illustration of propagation phenomena. He denied that the ether could have any detectable motion or momentum. As long as he believed in a strict validity of both the relativity principle and the reaction principle (for matter alone), he could not take the ether very seriously. However, at the beginning of the century he came to doubt the validity of the reaction principle (when applied to matter alone). The main source of this doubt was Kaufmann's experiments and their interpretation by electromagnetic inertia. Also, Poincaré was eager to correct excessive interpretations of the conventionalism he had earlier promoted:

> 'Did you not write,' you might say if you were seeking a quarrel with me, 'did you not write that the principles, though they are of experimental origin, are now beyond the possibility of experimental attack, because they have become conventions? And now you come to tell us that the triumphs of the most recent experiments put these principles in danger.' Very well, I was right formerly, and I am not wrong today. I was right formerly, and what is taking place at present is another proof of it.

In a hardened conventionalism, the necessity of certain conventions and the superfluity of others are exaggerated. In 1902, in a refutation of Edouard Le Roy's nominalism, Poincaré defended the ether against the latter sort of exaggeration:

> It can be said, for instance, that the ether has less reality than any external body. To say that this body exists is to say that there is an intimate, robust, and persistent relation between its color, its flavor, and its odor. To say that the ether exists, is to say that there is a natural relationship between all kinds of optical phenomena. Evidently, one proposition does not weigh more than the other.

Poincaré had offered the same comparison in 1888, with the nearly opposite conclusion that the ether, unlike ordinary bodies, would probably someday be rejected.[72]

[72]Poincaré, ref. 44, 322 [N'avez-vous pas écrit, pourriez-vous me dire si vous vouliez me chercher querelle, n'avez-vous pas écrit que les principes, quoique d'origine expérimentale, sont maintenant hors des atteintes de l'expérience parce qu'ils sont devenus des conventions? Et maintenant vous venez nous dire que les conquêtes les plus récentes de l'expérience mettent ces principes en danger. Et bien, j'avais raison autrefois et je n'ai pas tort aujourd'hui.]; "Sur la valeur objective de la science," *Revue de métaphysique et de morale*, 10 (1902), 263–293, on 293 [On peut dire par exemple que l'éther n'a pas moins de réalité qu'un corps extérieur quelconque ; dire que ce corps existe, c'est dire qu'il y a entre la couleur de ce corps, sa saveur, son odeur, un lien intime, solide

After Kaufmann's experiments of 1901, Poincaré admitted the possibility that the ether could carry momentum and even that all matter could be represented as a set of singularities of the ether. As we saw, in 1904 he contemplated the possibility that any inertia would be of electromagnetic origin, in which case the ether would carry every momentum in nature. In 1906, he cautiously announced "the end of matter" in a popular conference:

> One of the most astonishing discoveries announced by physicists during the past few years is that matter does not exist. I hurry to say that this discovery is not yet definitive... [If Kaufmann's results and the Lorentz–Abraham theory are correct], every atom of matter would be made of positive electrons, small and heavy, and of negative electrons, big and light... Both kinds have no mass and they only have a borrowed inertia. In that system, there is no true matter; there are only holes in the ether.

One might think that the expression "hole in the ether," which Poincaré repeated in later conferences, was meant to capture the imagination of a popular audience. This would however contradict Poincaré's understanding of the purpose of popularization, which we may infer from his appreciation of Kelvin's *Popular lectures*:

> Another remark immediately comes to mind. Where should we search for [Kelvin's] deepest ideas? In his Popular lectures. These lectures are not mere popularizations for which he would have more or less reluctantly sacrificed a few hours taken from more serious work. He did not humble himself in speaking to the people, for it is often in front of them and for them that his thoughts arose and took their most original form. Therefore, the same pages offer substance both to the beginner and to the scholar. How so? The evident reason is the nature of his mind: he did not think in formulas, he thought in images. The presence of a popular audience, the necessity to be understood from this audience naturally suggested images to him, images which were the normal generator of his thinking.

Poincaré, like Kelvin, liked to think through images, and some of his best ideas emerged when he was trying to find the right image. His seemingly naïve description of the ether reflects a move toward a more physical concept of the ether as a momentum carrier. Late in his life in a conference on materialism, he insisted: "Thus the active role is removed from matter to be transferred to the ether, which is the true seat of the phenomena that we attribute to mass. Matter no longer is; there are only holes in the ether."[73]

---

et persistant ; dire que l'éther existe, c'est dire qu'il y a une parenté naturelle entre tous les phénomènes optiques, et les deux propositions n'ont évidemment pas moins de valeur l'une que l'autre.].

[73]Poincaré, "La fin de la matière," *Athenaeum*, 4086 (1906), 201–202 [L'une des découvertes les plus étonnantes que les physiciens aient annoncées dans ces dernières années, c'est que la matière n'existe pas. Hâtons-nous de dire que cette découverte n'est pas encore définitive... Ainsi tout atome matériel serait formé d'électrons positifs, petits et lourds, et d'électrons négatifs, gros et

To summarize, after Kaufmann's experimental discovery of electromagnetic inertia and after his own discovery that moving observers measure Lorentz-transformed fields and coordinates, Poincaré employed the ether in two manner: as a momentum carrier in which momentum was globally conserved, and as a privileged reference frame in which the conventions of the usual geometry could be maintained. Whereas for Einstein the ether became superfluous in relativity theory, for Poincaré the ether offered a way to conciliate relativity theory with older intuitions and conventions. This is not to say that Poincaré was unaware of the possibility of changing the conventions to get rid of the ether. He just did not believe in the necessity of such a radical move.

Conclusions

The physics of light inspired Poincaré in four different manners: as a useful employment of his mathematical skills, as an entry into methodology and philosophy, as a touchstone for electromagnetic theory, and as a precondition for a new light-based metrology. In the first register, he greatly advanced diffraction theory both in the optical and in the Hertzian domain. In the second, he developed a pluralist view of physics in which several illustrative models were allowed to coexist even though the shared structure was the essential, stable core of the theory; he detected and defended a move of physics from specific modeling to organizing principles; he articulated a nuanced understanding of the principles of physics as provisionally convenient conventions of empirical origin. In the third register, he discussed the compatibility of the optics of moving bodies with the electromagnetic interpretation of light; this led him to the first general formulation of the relativity principle and to the full group-theoretical apparatus of relativity theory. In the fourth and last register, he recognized that the time coordinate of optically synchronized clocks should depend on the motion of the reference frame to which they are attached, and that optically measured lengths were Lorentz-contracted.

---

légers... Les uns et les autres sont dépourvus de masse et n'ont qu'une inertie d'emprunt. Dans ce système il n'y a pas de vraie matière. Il n'y a plus que des trous dans l'éther] (Poincaré was aware of new, contradictory experiments of Kaufmann, but judged that a definitive conclusion was premature); "Lord Kelvin," *La lumière électrique*, 1 (1908), 139–147, on 139 [On ne peut s'empêcher de faire une autre remarque. Où faut-il chercher ses idées les plus profondes? Dans ses *Popular Lectures*. Ces leçons ne sont donc pas de simples vulgarisations, en vue desquelles il aurait sacrifié plus ou moins à regret quelques heures prises sur un travail plus sérieux. Il ne s'abaissait pas pour parler au peuple, puisque c'est souvent devant et pour lui que sa pensée prenait naissance et revêtait sa forme la plus originale. C'est donc dans les mêmes pages que le lecteur novice et le savant pourront aller chercher et trouver un aliment. Comment cela se fait-il? Cela vient évidemment de la nature de son esprit, il ne pensait pas en formules, il pensait en images ; la présence de l'auditoire populaire, la nécessité de s'en faire comprendre lui suggérait naturellement l'image, qui était pour lui la génératrice habituelle de la pensée]; "Les conceptions nouvelles de la matière," in Paul Doumergue (ed.), *Le matérialisme actuel* (Paris, 1913), 49–67, on 65 [Voilà le rôle actif enlevé à la matière, pour être transféré à l'éther, véritable siège des phénomènes que nous attribuons à la masse. Il n'y a plus de matière, il n'y a plus que des trous dans l'éther.]

These achievements of Poincaré were deeply interrelated: the mathematical development of various ether theories helped him identify their shared mathematical structure; considerations of structure and principles inspired his critical insights into the optics and electrodynamics of moving bodies; his awareness of a new light-based metrology arose in this context. His concern with light, its behavior, its nature, and its uses thus offers a unified explanation of its most important breakthroughs in mathematical physics.

Appendix – Light-based measurement in the Lorentz–Poincaré ether theory

Consider a rod AB moving uniformly in the ether at velocity u along the x axis and making a constant angle with this axis (the rod could be one arm of a Michelson interferometer, or a materialization of the distance between two terrestrial observers). A flash of light is emitted at one end of the rod; it is reflected at the other extremity, and returns to the first end. Call x the projection of the length of the rod on the x axis, y the projection on the perpendicular axis in the plane of the rod, t_+ the time that the light takes to travel from A to B, and t_- the time it takes to travel from B to A. Drawing the light trajectories in the ether frame as it done for t_+ in Fig. 7 (Poincaré imagines such figures for the Michelson–Morley experiment and for the first-order local time), we have

$$c^2t_+^2 = y^2 + (x + ut_+)^2 \ , \quad c^2t_-^2 = y^2 + (x - ut_-)^2. \tag{44}$$

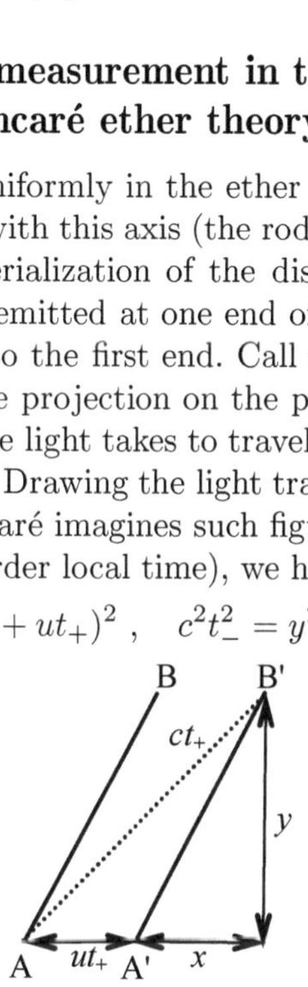

FIGURE 7. The path (AB′) of light between the extremities of a rod AB moving at the constant velocity u.

The difference of these two equations leads to

$$t_+ - t_- = 2ux/(c^2 - u^2) \ . \tag{45}$$

According to Poincaré's optical synchronization, two events occurring in A and B are judged simultaneous in the moving frame if and only if the difference of their true times is $(t_+ - t_-)/2$. Now suppose that x is proportional to the difference $x'_A - x'_B$ of the apparent abscissae of A and B. Then apparent simultaneity is transitive because

$$x'_A - x'_C = (x'_A - x'_B) - (x'_C - x'_B) \ . \tag{46}$$

This is true whatever the contraction factor for the parallel component of the rod may be. Hence Poincaré erred when, in 1906–1907, he claimed that the Lorentz value of the contraction was necessary to warrant the transitivity of apparent synchronization. Any contraction or no contraction at all would do.

However, the Lorentz value is necessary in order that the traveling times be the same in the two arms of a Michelson interferometer. That the Lorentz contraction still does the job for any orientation of the interferometer or for any angle between its two arms can be proved by showing that the round-trip time $t_+ + t_-$ does not depend on the orientation of the rod AB if we assume the Lorentz contraction

$$x = x'\sqrt{1 - u^2/c^2}, y = y' \,, \tag{47}$$

x' and y' being the two projections of the rod when it does not move (or the apparent projections in the moving frame). Adding the two equations (44) and using equation (45), we get

$$(t_+ + t_-)^2 = \tfrac{4}{c^2-u^2}\left(\tfrac{x^2}{1-u^2/c^2} + y^2\right). \tag{48}$$

Hence the Lorentz contraction leads to

$$(t_+ + t_-)^2 = \tfrac{4}{c^2-u^2}(x'^2 + y'^2) = \tfrac{4l^2}{c^2-u^2} \,, \tag{49}$$

if l denotes the length of the rod at rest. This quantity does not depend on the inclination of the rod, as was to be proved.

Possibly, sometime after showing the ellipsoid to his students Poincaré realized that the Lorentz contraction was not required for the transitivity of optical synchronization. In the 1908 version of the ellipsoid, transitivity is no longer mentioned; the new purpose of the ellipsoid is to show that "the *apparent* duration of transmission is proportional to the *apparent* distance," with the comment: "This time the compensation is *rigorous*, and this is the explanation of Michelson's experiment." Indeed, in the ellipsoid argument, it is only for Lorentz's value of the contraction that the focus F of the ellipse coincides with the apparent position of the source. And it is only in this case that FM is proportional to the apparent transmission time.

Possibly, Poincaré arrived at the light ellipsoid by geometrically interpreting the equations (44). Indeed, in terms of the non-contracted projections x' and y' given by (47), we have

$$t_+^2 = y'^2 + (\gamma^{-1}x' + ut_+)^2, \tag{50}$$

which is the equation of an elongated ellipsoid of eccentricity u/c if $\gamma = \frac{1}{\sqrt{1-u^2/c^2}}$, with the focus F such that $\mathrm{OF} = \gamma u t_+$ (and therefore coinciding with the apparent position of the source). This is exactly the Poincaré ellipsoid.

Acknowledgments

I thank Thibault Damour and Bertrand Duplantier for their useful comments on a draft of this paper, and Chantal Delongeas for producing the printable version.

Olivier Darrigol
CNRS: Laboratoire SΦHÈRE, UMR 319
e-mail: `darrigol@paris7.jussieu.fr`

Henri Poincaré, 1912–2012, 51–149

Poincaré Seminar 2012

Poincaré and the Three-Body Problem

Alain Chenciner

Abstract. The Three-Body Problem has been a recurrent theme of Poincaré's thought. Having understood very early the need for a qualitative study of "non-integrable" differential equations, he developed the necessary fundamental tools: analysis, of course, but also topology, geometry, probability. One century later, mathematicians working on the Three-Body Problem still draw inspiration from his works, in particular in the three volumes of *Les méthodes nouvelles de la mécanique céleste* published respectively in 1892, 1893, 1899.

1. Introduction

Since the time of Newton himself, the Three-Body Problem was a major source of development of analysis: it is enough to mention the names of Euler, Clairaut, d'Alembert, Laplace, Lagrange, Jacobi, Cauchy, ... At the end of the nineteenth century, Poincaré opened a new era, introducing geometric, topological and probabilistic methods in order to understand qualitatively the incredibly complicated behavior of most of the solutions of this problem. At the same time, he analysed the methods used by the astronomers in order to understand the short-term motions using divergent series and, as emphasized in a recent paper by J.P. Ramis [Ra1], he foresaw many aspects for the present development of this theme. From 1883, that is only 4 years after his thesis defense, and until his death in 1912, Poincaré published major papers on (or motivated by) the Three-Body Problem. Already in 1881, in the introduction to the first part of his *Mémoire sur les courbes définies par une équation différentielle* [P3], he took it as motivation for a qualitative global study:

> Prenons, par exemple, le problème des trois corps : ne peut-on pas se demander si l'un des corps restera toujours dans une certaine région du ciel ou bien s'il pourra s'éloigner indéfiniment ; si la distance de deux corps augmentera, ou diminuera à l'infini, ou bien si elle restera comprise entre certaines limites ? Ne peut-on pas se poser mille questions de ce genre, qui seront toutes résolues quand on saura construire qualitativement les

> trajectoires des trois corps ? Et, si l'on considère un nombre plus grand de corps, qu'est-ce que la question de l'invariabilité des éléments des planètes, sinon une véritable question de géométrie qualitative, puisque, faire voir que le grand axe n'a pas de variations séculaires, c'est montrer qu'il oscille constamment entre certaines limites. Tel est le vaste champ de découvertes qui s'ouvre devant les géomètres.[1]

In 1885, introducing the third part of the same *Mémoire*, he was more precise and addressed the reader in his characteristic style:

> On n'a pu lire les deux premières parties de ce Mémoire sans être frappé de la ressemblance que présentent les diverses questions qui y sont traitées avec le grand problème astronomique de la stabilité du système solaire. Ce dernier problème est, bien entendu, beaucoup plus compliqué, puisque les équations différentielles du mouvement des corps célestes sont d'ordre très élevé. Il y a même plus, on rencontrera, dans ce problème, une difficulté nouvelle, essentiellement différente de celles que nous avons eu à surmonter dans l'étude du premier ordre, et j'ai l'intention de la faire ressortir, sinon dans cette troisième Partie, du moins dans la suite de ce travail.[2]

The difficulty alluded to by Poincaré is caused by the so-called *small denominators* which appear in the perturbation series of the astronomers; and indeed, having devoted the end of the *Mémoire* to the study of the *Lindstedt series* which govern the problematic existence of families of quasi-periodic solutions[3] "surrounding" a periodic one in $\mathbb{R}^3$, he concluded as follows:

> D'après ce qui précède, on comprendra sans peine à quel point les difficultés que l'on rencontre en Mécanique céleste, par suite des petits diviseurs et de la quasi-commensurabilité des moyens mouvements, tiennent à la nature même des choses et ne peuvent être tournées. Il est

[1]Let us take, for example, the Three-Body Problem: is it not possible to ask whether one of the bodies will remain forever in some region of the sky or whether it will possibly get away indefinitely; whether the distance between two bodies will increase, or decrease indefinitely, or whether it will stay bounded between some limits? Is it not possible to ask a thousand similar questions, which will all be solved as soon as one is able to construct qualitatively the trajectories of the three bodies? And if one considers more bodies, what is the question of the invariability of the elements of the planets but a true question of qualitative geometry, as showing that the great axis has no secular variations amounts to showing that it oscillates between some limits. Such is the vast field of discoveries which opens up to geometers.

[2]One cannot have read the first two parts of this Memoir without having been struck by the similarities of the various questions studied there with the great astronomical problem of the stability of the solar system. This last problem is, of course, much more complicated, because the differential equations of motion of the celestial bodies are of a very high order. What is more, one shall encounter in this problem a new difficulty, which differs essentially from the ones we had to overcome in the first-order study, and I intend to bring it out, if not in this third Part, at least in the remainder of this work.

[3]More precisely, families of invariant tori defined by a conserved quantity.

> extrêmement probable qu'on les retrouvera, quelle que soit la méthode que l'on emploie.[4]

A landmark in Poincaré's works on the Three-Body Problem is the famous Memoir *Sur le problème des trois corps et les équations de la dynamique* [P1], winner in 1889 of the prize given on the occasion of the 60th birthday of the King of Sweden, and even more *Les méthodes nouvelles de la mécanique céleste* [P2] whose three volumes, totaling almost 1300 pages, appear respectively in 1892, 1893 and 1899. Vastly enlarging the scope of the Memoir, this extraordinary work, which encompasses and develops most of Poincaré's previous researches on the Three-Body Problem, is the source of a major part of the modern theory of *Dynamical Systems: normal forms, exponents, invariant manifolds, homoclinic and heteroclinic solutions, analytic non-integrability, divergence of the perturbation series and exponentially small splitting of separatrices, variational equations and integral invariants, generating functions, recurrence theorem, surfaces of section and return maps, twisting property,* all of them are part of the present landscape and they paved the way for *bifurcation studies and the theory of singularities, symbolic dynamics, invariant measures and ergodic theory, K.A.M., weak K.A.M. and diffusion, symplectic geometry* ... and also a wealth of computer experiments. Indeed, in the realm of Dynamical Systems, "new" ideas which are not in one way or another rooted in Poincaré's works are few and far between ([AKN, HK, C0]). *Geometry* is a key word but it is amusing to remark how cautiously Poincaré mentions its use in higher dimensions at the beginning of the last part of the *Mémoire* of 1886 [P3]:

> Si l'on veut, dans le cas des équations (2), employer le mode de représentation géométrique dont nous avons fait usage jusqu'ici, il faut regarder $x_1, x_2, \ldots, x_n$ comme les coordonnées d'un point dans l'espace à n dimensions.
>
> La Géométrie n'est plus alors qu'un langage qui peut être plus ou moins avantageux, ce n'est plus une représentation parlant aux sens. Nous pourrons néanmoins être conduits à employer quelquefois ce langage.[5]

Most of what follows will be dedicated to *Les méthodes nouvelles de la mécanique céleste*,[6] that is to the perturbative case where a small parameter (of the nature of

[4]According to what was said above, one will readily appreciate to what extent the difficulties encountered in Celestial Mechanics because of the small denominators and the quasi-commensurability of the mean motions, are attached to the very nature of things and cannot be avoided. It is highly probable that they will come up whatever method we use.

[5]If one wants to use for equations (2) the geometric way of representation which we have used up to now, one must consider $x_1, x_2, \ldots, x_n$ as the coordinates of a point in an n-dimensional space. In this case, Geometry is no more than a language that may be more or less advantageous; it is no longer a representation addressing the senses. Nonetheless, we will occasionally be led to employ this language.

[6]From now on, quoted as *The New Methods*.

the problem is naturally studied), we get[12]

$$\frac{dp_k}{dt} = -\frac{\partial F}{\partial q_k}, \quad \frac{dq_k}{dt} = \frac{\partial F}{\partial p_k}, \quad \text{where} \quad F = \sum_i \frac{1}{2m_i}||\vec{\pi}_i||^2 - \sum_{i<j} \frac{m_i m_j}{||\vec{r}_j - \vec{r}_i||}.$$

(Beware that the index k takes values from 1 to 9 if we are looking at the Three-Body Problem in $\mathbb{R}^3$.)

The simplest case of interest in classical astronomy is the *planetary problem*[13], or $1+2$ *body problem*, where two planets revolve around a much more massive Sun (a mass ratio $m_i/m_0 = 1/1000$ in the case of Jupiter). If moreover, as we shall suppose, the two orbits of the planets stay far apart[14], the equations may be seen as a small perturbation of the limit equations where each planet revolves around the Sun along a Keplerian ellipse unperturbed by the other one. Thus, the first volume of *The New Methods* starts by recalling the solution of the *Two-Body Problem* or, what amounts to the same, the *Kepler Problem* or attraction by a fixed center.

2.2. Kepler Problem: Delaunay and Poincaré coordinates

When there are only two bodies, the relative position vector $\vec{r} = \vec{r}_1 - \vec{r}_0$ satisfies the so-called *Kepler equation*[15]

$$\frac{d^2\vec{r}}{dt^2} = -\frac{(m_0+m_1)\vec{r}}{||\vec{r}||^3}, \quad \text{with Hamiltonian} \quad h = \frac{1}{2m}||\vec{\pi}||^2 - \frac{mM}{||\vec{r}||},$$

where we have denoted $M = m_0 + m_1$ and the choice of the *reduced mass* $m = \frac{m_0 m_1}{m_0+m_1}$ leaves the potential unchanged. We are only interested here in elliptic solutions, that is motions with negative energy h. Introducing *Delaunay coordinates*

$$x = (L, G, \Theta),\ y = (l, g, \theta)$$

(in the case of the problem in $\mathbb{R}^3$, see Figure 1 and [S1] for details[16]) the (still canonical) equations become simply

$$\frac{dx_k}{dt} = -\frac{\partial F}{\partial y_k}, \quad \frac{dy_k}{dt} = \frac{\partial F}{\partial x_k}, \quad \text{where} \quad F = F(x) = -\frac{m^3M^2}{2L^2}.$$

Thus the only non-constant quantity is l which varies proportionally to the time and one recovers the three Kepler laws. That the Hamiltonian does not depend on the *angles* $y = (l, g, \theta)$ but only on the *actions* $x = (L, G, \Theta)$ means that the Delaunay variables are *action-angle* coordinates for the Kepler Problem; that it does

---

[12]The sign convention is not Poincaré's own in his statement of the general problem of dynamics (see 2.5; actually, Poincaré is not consistent in his naming of the variables and hence in his choices of signs: denoting by (x, y) the coordinates in phase space, the actions are sometimes represented by x and sometimes by y).

[13]There is also the *lunar problem*, to which I shall allude when discussing the *Restricted Problem*.

[14]This excludes in particular the Lagrange equilateral solutions for which the limit orbits share the same value of the semi-major axis.

[15]which, of course, Kepler never wrote.

[16]Neither the sign conventions nor the constants are the same as those of Poincaré who worked in conformal symplectic coordinates and not in symplectic ones.

not depend on G and Θ reflects the degeneracy of the Kepler Problem: the angles g and θ stay constant, which means that the ellipse along which the motion takes place has no precession (according to *Bertrand's theorem* this characterizes Newton's equations and the harmonic oscillator equations among all laws of attraction which are powers of the distance and even all central force laws).

$L = \sqrt{-\frac{m^3M^2}{2h}} = m\sqrt{M}\sqrt{a}$

h = energy, a = semi-major axis

$G = m\sqrt{M}\sqrt{a(1-e^2)} = |\vec{C}|$

$\vec{C}$ = angular momentum,

e = eccentricity

$\Theta = G\cos i$,

i = inclination

l = mean anomaly

= proportion of area swept

g = argument of perihelion,

θ = longitude of the node

$n = \frac{dl}{dt}$, $n^2a^3 = M$.

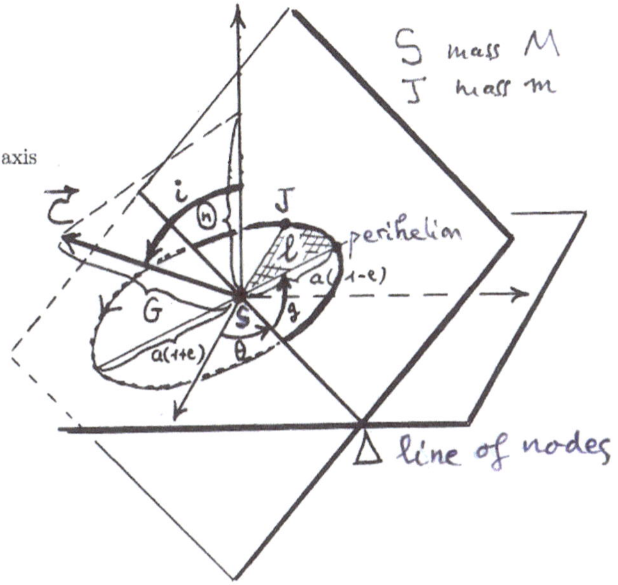

FIGURE 1. Delaunay coordinates. Reproduced from [S1] with permission of © IMCCE 1989.

Delaunay coordinates have a major defect for the planetary problem: they do not extend to Keplerian orbits whose eccentricity and inclination vanish. This led Poincaré to introduce a symplectic system of coordinates $\big((\Lambda,\lambda),(\xi,\eta),(p,q)\big)$, which is obtained from the Delaunay coordinates by, first, making a linear change of coordinates:

$$\begin{cases} \Lambda = L, \quad H = L - G, \quad Z = G - \Theta, \\ \lambda = l + g + \theta, \quad h = -g - \theta, \quad \zeta = -\theta. \end{cases}$$

and then considering the pairs (H,h) et (Z,ζ) as planar symplectic polar coordinates:

$$\begin{cases} \sqrt{2H}\cos h = \xi, \quad \sqrt{2H}\sin h = \eta, \\ \sqrt{2Z}\cos\zeta = p, \quad \sqrt{2Z}\sin\zeta = q. \end{cases}$$

Called today *Poincaré coordinates*[17], these coordinates are pertinent in the neighborhood of zero eccentricity ($H = 0$) or zero inclination ($Z = 0$)[18]. The conjugate variables (Λ, λ) respectively describe the semi-major axis (i.e., the energy) and the *mean longitude* (i.e., the position of the body on its elliptic trajectory), while the conjugate variables (ξ, η) describe the position of the ellipse within its plane and its eccentricity, and the set of conjugate variables (p, q) (not to be confused with the position and momentum coordinates introduced in 2.1) describe the position of the plane containing the ellipse and its inclination with respect to the reference plane.

2.3. Planetary problem: heliocentric coordinates

Reducing the Galilean symmetry by the use of Jacobi coordinates (as Poincaré did in the first chapter of *The New Methods*) or canonical heliocentric coordinates (that he will introduce only in 1896, see [S1] for a detailed study of both) one writes the planetary problem as a perturbation of two uncoupled fictitious Kepler Problems.

Let us choose for instance the canonical heliocentric coordinates:

$$\begin{cases} \vec{Q}_0 = \vec{r}_0, & \vec{Q}_i = \vec{r}_i - \vec{r}_0,\ i = 1, 2, \\ \vec{P}_0 = \vec{\pi}_0 + \vec{\pi}_1 + \vec{\pi}_2, & \vec{P}_i = \vec{\pi}_i,\ i = 1, 2. \end{cases}$$

The Hamiltonian

$$\sum_{i=0}^{2} \frac{|p_i|^2}{2m_i} - \sum_{0 \le i < j \le 2} \frac{m_i m_j}{|q_j - q_i|}$$

of the $1 + 2$ body problem becomes (after setting $P_0 = 0$, which amounts to fixing the center of mass)

$$\begin{cases} \mathcal{H}(\vec{P}_1, \vec{P}_2, \vec{Q}_1, \vec{Q}_2) = \mathcal{H}(\vec{P}, \vec{Q}) = \mathcal{H}_0(\vec{P}, \vec{Q}) + \mathcal{H}_1(\vec{P}, \vec{Q}), \\ \mathcal{H}_0 = \sum_{i=1}^{2} |\vec{P}_i|^2 \left(\frac{1}{2m_0} + \frac{1}{2m_i} \right) - \sum_{i=1}^{2} \frac{m_0 m_i}{|\vec{Q}_i|}, \\ \mathcal{H}_1 = \mathcal{H}_{1p} + \mathcal{H}_{1c} = -\frac{m_1 m_2}{|\vec{Q}_2 - \vec{Q}_1|} + \frac{\vec{P}_1 \cdot \vec{P}_2}{m_0}. \end{cases}$$

The equations (written in a concise way which, as Poincaré would say, “ne pouvait d’ailleurs tromper personne”[19])

$$\frac{d\vec{P}_i}{dt} = -\frac{\partial \mathcal{H}}{\partial \vec{Q}_i}, \quad \frac{d\vec{Q}_i}{dt} = \frac{\partial \mathcal{H}}{\partial \vec{P}_i},$$

---

[17] The coordinates we have defined are adapted to solutions which turn in the positive, prograde, direction. One can introduce in the same way Poincaré coordinates adapted to retrograde motions.

[18] In the first volume of the *Lessons* [P12], Poincaré gave a detailed proof of the analyticity of these coordinates.

[19] could not anyway mislead anyone (compare with the beginning of Section 341 of *The New Methods*).

may be interpreted as describing the motion of two fictitious planets around a fixed center. The motions corresponding to the unperturbed Hamiltonian $\mathcal{H}_0$ are those of two uncoupled Kepler Problems for which the mass of the fixed center and the mass of the planet are respectively

$$\bar{M}_i = m_0 + m_i \text{ and } \bar{m}_i = \frac{m_0 m_i}{m_0 + m_i}, \quad i = 1, 2.$$

The perturbing function is the sum of a *principal part* $\mathcal{H}_{1p}$ and a *complementary part* $\mathcal{H}_{1c}$. It remains small as long as m_0 is much larger than m_1, m_2 and the two unperturbed ellipses are well separated.

Recall the *conservation of angular momentum*[20]:

$$\vec{C} = \sum_{i=0}^{2} m_i \vec{r}_i \wedge \frac{d\vec{r}_i}{dt} = \sum_{i=0}^{2} \vec{r}_i \wedge \vec{\pi}_i = \sum_{i=0}^{2} \vec{Q}_i \wedge \vec{P}_i = \sum_{i=1}^{2} \vec{Q}_i \wedge \vec{P}_i.$$

In $\mathbb{R}^3$, the (invariant) plane orthogonal to $\vec{C}$ (when $\vec{C} \neq 0$) is called the *Laplace plane*. The expression of the angular momentum as the sum of the angular momenta of the two fictitious bodies with position $\vec{Q}_i$ and momentum $\vec{P}_i$, $i = 1, 2$, shows that the Laplace plane and the two planes spanned respectively by $(\vec{Q}_1, \vec{P}_1)$ and $(\vec{Q}_2, \vec{P}_2)$ have a common intersection, called the line of nodes as in Figure 2. We may now introduce Delaunay coordinates $(L_1, G_1, \Theta_1, l_1, g_1, \theta_1)$, $(L_2, G_2, \Theta_2, l_2, g_2, \theta_2)$, for each one of the Kepler Problems[21]. As Poincaré explains in the first chapter of [P2][22]:

> Si alors, à un instant quelconque, les forces appliquées à la première masse fictive venaient à disparaître, et qu'elles soient remplacées par l'attraction d'une masse m_0+m_1 placée à l'origine, cette masse se mouvrait suivant les lois de Képler et les éléments de ce mouvement képlérien seraient L, G, Θ, l, g et θ.[23]

Warning! We shall choose, as is usual (see for instance [Ti, Cha]), to define θ_j as the longitude of the ascending node of the jth planet. This implies $\theta_2 - \theta_1 = \pi$ mod 2π. On the contrary, in Section 16, Poincaré writes that $\theta_1 = \theta_2$ without making explicit the fact that there is a choice.

Setting

$$x = (L_1, G_1, \Theta_1, L_2, G_2, \Theta_2), \quad y = (l_1, g_1, \theta_1, l_2, g_2, \theta_2),$$

---

[20] The angular momentum is a bivector but, in $\mathbb{R}^3$ (resp. $\mathbb{R}^2$) endowed with its canonical orientation and canonical Euclidean structure, it may be identified with a vector (resp. a real number).

[21] which give slightly different masses $\tilde{M}_1$ and $\tilde{M}_2$ to the Sun.

[22] Since Poincaré used Jacobi coordinates and numbers 1,2,3 the bodies, I needed to change the mass of the Sun in the quotation which originally had $m_1 + m_2$. Astronomers used to dislike heliocentric coordinates because the fictitious ellipses are not osculating the real trajectories, but this is not significant.

[23] If, at some instant, the forces which are applied to the first fictitious mass happened to vanish, and if they were replaced by the attraction of a mass $m_0 + m_1$ located at the origin, this mass would move according to Kepler laws and the elements of this Keplerian motion would be L, G, Θ, l, g and θ.

the equations are still canonical with a Hamiltonian of the form

$$\frac{K_1}{L_1^2} + \frac{K_2}{L_2^2} + R(x, y),$$

where K_1, K_2 are constants and the *perturbing function* R depends on all the variables.

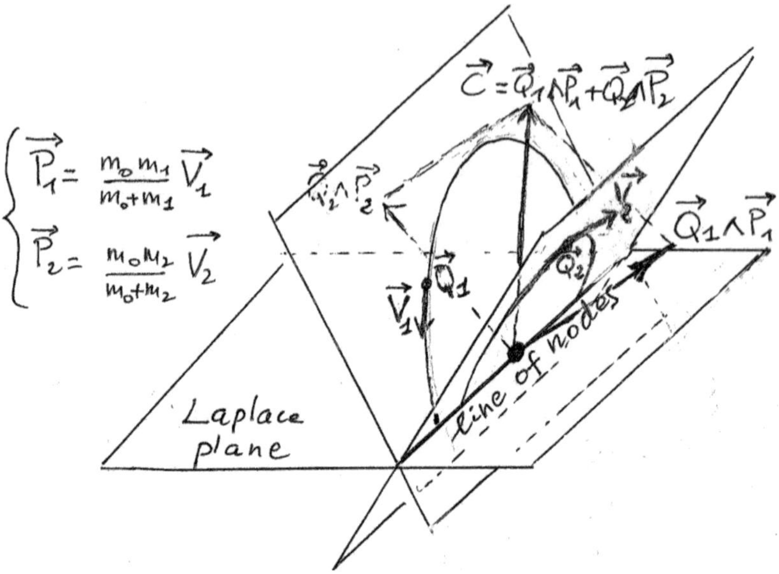

FIGURE 2. The fictitious Keplerian motions at some instant.

2.4. Reduced problem

When fixing the direction of the angular momentum to be vertical in the preceding section, we have already performed a *partial reduction*[24] of the rotational symmetry, which amounts to restricting the dynamics to the (symplectic) invariant submanifold of the phase space defined by

$$\theta_2 - \theta_1 = \pi, \quad G_2^2 - \Theta_2^2 = G_1^2 - \Theta_1^2.$$

Since fixing the norm of the angular momentum is equivalent to setting $\Theta_1 + \Theta_2 = C$, taking the time derivative, we get that $\frac{\partial F}{\partial \theta_1} + \frac{\partial F}{\partial \theta_2} = 0$, from which it follows that F depends only on the difference $\theta_2 - \theta_1$. Since this difference is constant, F does not depend on θ_1, θ_2 once the angular momentum is fixed, and we can perform a complete symplectic reduction by keeping only the 8 variables

$$L_1, L_2, l_1, l_2, G_1, G_2, g_1, g_2$$

[24]Partial reduction will play an important role in Section 11.3.

The three *area integrals*, that is the three components of the angular momentum, are still first integrals of the secular Hamiltonian. If one fixes the angular momentum in the direction of the vertical axis, they become

$$\begin{cases} \sum_{j=1}^{n}\left(\Lambda_j - \frac{1}{2}\left(|u_j|^2 + |z_j|^2\right)\right) = \text{ a positive real constant,} \\ \sum_{j=1}^{n} z_j\sqrt{2\Lambda_j - |u_j|^2 - \frac{1}{2}|z_j|^2} = 0 \in \mathbb{C}, \end{cases}$$

(the reader too lazy to do the computation may look at the first volume of the *Leçons de Mécanique Céleste* ([P12] Section 144).

The Λ_j being constants, the first equation defines a sphere S^7 in $\mathbb{R}^8 = \mathbb{C}^4$ surrounding the singularity at the origin, corresponding to a pair of horizontal circles; this is the second part of *Laplace stability theorem*: at the secular level, the eccentricities and inclinations remain small for all time if they are small initially. If the radius ϵ of this sphere is chosen sufficiently small, the submanifold J^5 of $\mathbb{R}^8$ defined by the three equations above is diffeomorphic to the intersection of S^7 with the complex hyperplane whose equation is $\sum_{i=1}^{2} z_i\sqrt{2\Lambda_i} = 0$, that is to S^5. A quick inspection of Poincaré coordinates shows that the group of rotations around Ox_3 acts diagonally on $\mathbb{R}^8 = \mathbb{C}^4$:

$$e^{i\alpha}\cdot(u_1, u_2, z_1, z_2) = \left(u_1e^{i\alpha}, u_2e^{i\alpha}, z_1e^{i\alpha}, z_2e^{i\alpha}\right).$$

Hence the quotient K^4 of J^5 by this action (which preserves it) is diffeomorphic to the quotient of the round sphere S^5, that is to the complex projective space $P^2(\mathbb{C})$. Note that K^4 is naturally a submanifold of the quotient $P^3(\mathbb{C})$ of S^7 on which one can use the homogeneous complex coordinates $u_1 : u_2 : z_1 : z_2$, and that the invariant submanifold defined by the motions in the horizontal plane coincides with the subspace $P^1(\mathbb{C})$ (= the *Riemann sphere*) obtained by setting $z_1 = z_2 = 0$ (and on which $u_1 : u_2$ are homogeneous coordinates). Finally, after reduction of the rotation symmetry, the secular spatial (resp. planar) $1+2$ body problem amounts to the study of a Hamiltonian system on a submanifold K^4 (resp. K^2) of $P^3(\mathbb{C})$ (resp. $P^2(\mathbb{C})$) diffeomorphic to $P^2(\mathbb{C})$ (resp. $P^1(\mathbb{C})$). The complement in K^4 of the $P^1(\mathbb{C})$ corresponding to the horizontal motions is diffeomorphic to the complement of $P^1(\mathbb{C})$ in $P^2(\mathbb{C})$, that is $\mathbb{C}^2 = \mathbb{R}^4$, and the mapping

$$D_\epsilon = \left\{(u_1, u_2) \in \mathbb{C}^2,\ |u_1|^2 + |u_2|^2 < \epsilon^2\right\} \ni (u_1, u_2) \mapsto (u_1, u_2, z_1, z_2) \in J^5$$

uniquely defined by the conditions that z_1 et z_2 be real and $z_1 > 0$ gives in the quotient a symplectic diffeomorphism of D_ϵ endowed with the symplectic form $du_1 \wedge d\bar{u}_1 + du_2 \wedge d\bar{u}_2$ onto $K^4 \setminus P^1(\mathbb{C})$ endowed with the symplectic form coming from the reduction. Writing the secular Hamiltonian in such coordinates amounts to performing Jacobi's *reduction of the node*: geometrically, it corresponds to the identity $\theta_2 - \theta_1 = \pi$, which is satisfied when the angular momentum is vertical (compare Figure 2). This identity implies the disappearance of the angles θ_1 et

θ_2 from the expression of the Hamiltonian which, because of its invariance under rotations around the vertical axis, depends only on their difference.

3.3. Quadratic part and singularities

Once the angular momentum has been fixed in the vertical direction, the secular Hamiltonian admits two symmetries in addition to the rotations around the vertical axis:

1) the symmetry with respect to a vertical plane, for example the plane $x_1 0 x_3$, which acts on the longitudes by

$$(g_1, g_2, \theta_1, \theta_2) \mapsto (\pi - g_1, \pi - g_2, \pi - \theta_1, \pi - \theta_2),$$

and becomes in the u, z coordinates

$$(u_1, u_2, z_1, z_2) \mapsto (\bar{u}_1, \bar{u}_2, -\bar{z}_1, -\bar{z}_2);$$

2) the rotation by π around the common node, which acts on the longitudes by

$$(g_1, g_2, \theta_1, \theta_2) \mapsto (-g_1, -g_2, \theta_1, \theta_2),$$

and becomes, after composition with the rotation by $-2\theta_1$ around the vertical axis,

$$(u_1, u_2, z_1, z_2) \mapsto (\bar{u}_1, \bar{u}_2, \bar{z}_1, \bar{z}_2).$$

Composing both and then rotating by π around the vertical axis, we get that the secular Hamiltonian is invariant by complex conjugation and by the transformation

$$(u_1, u_2, z_1, z_2) \mapsto (-u_1, -u_2, z_1, z_2).$$

It follows that the quadratic part of the Taylor expansion of the secular Hamiltonian at its "circular horizontal" singularity has the form

$$a_1|u_1|^2 + a_2|u_2|^2 + b\mathrm{Re}\,(u_1\bar{u}_2) + c_1|z_1|^2 + c_2|z_2|^2 + d\mathrm{Re}\,(z_1\bar{z}_2),$$

which one writes classically

$$\begin{aligned} Q(\xi_1, \xi_2, \eta_1, \eta_2, p_1, p_2, q_1, q_2) &= Q'(\xi_1, \xi_2) + Q'(\eta_1, \eta_2) \\ &\quad - Q''(p_1, p_2) - Q''(q_1, q_2), \end{aligned}$$

where Q' et Q'' are real quadratic forms of two variables known since the 18th century (see [Ti, Cha]):

$$\begin{cases} Q'(X,Y) = m_1 m_2 \left\{ \dfrac{1}{8} B^{(1)} \left(\dfrac{X^2}{\Lambda_1} + \dfrac{Y^2}{\Lambda_2} \right) - \dfrac{1}{4} B^{(2)} \dfrac{XY}{\sqrt{\Lambda_1 \Lambda_2}} \right\}, \\[2ex] Q''(X,Y) = m_1 m_2 \left\{ \dfrac{1}{8} B^{(1)} \left(\dfrac{X}{\sqrt{\Lambda_1}} - \dfrac{Y}{\sqrt{\Lambda_2}} \right)^2 \right\}, \end{cases}$$

where the *Laplace coefficients* $B^{(1)}$ and $B^{(1)}$ depend only on Λ_1, Λ_2. Keeping only the quadratic terms in the eccentricities e_1, e_2 and the inclinations i_1, i_2, and taking into account that $\theta_2 - \theta_1 = \pi$ (see 2.3), one finds that the dominant terms of the secular Hamiltonian are[30]:

$$m_1 m_2 \left[\frac{1}{8}B^{(1)}\left[e_1^2 + e_2^2 - (i_1 + i_2)^2\right] + \frac{1}{4}B^{(2)} e_1 e_2 \cos(g_1 - g_2)\right].$$

Finally, replacing the two equations which fix the direction of the angular momentum by their linearization[31] $z_1\sqrt{2\Lambda_1} + z_2\sqrt{2\Lambda_2} = 0$, and choosing in the subspace (diffeomorphic to $\mathbb{C}^3 = \mathbb{R}^6$) which they define the coordinates u_1, u_2, z_1, the restriction of Q becomes the quadratic form

$$\begin{cases} Q_0 = Q'(\xi_1, \xi_2) + Q'(\eta_1, \eta_2) - m_1 m_2 \dfrac{B^{(1)}\Lambda_1}{8}\left(\dfrac{1}{\Lambda_1} + \dfrac{1}{\Lambda_2}\right)^2 (p_1^2 + q_1^2) \\ \qquad = \dfrac{m_1 m_2}{8}\left\{B^{(1)}\left(\dfrac{|u_1|^2}{\Lambda_1} + \dfrac{|u_2|^2}{\Lambda_2}\right) - 2B^{(2)}\dfrac{\mathrm{Re}(u_1\bar{u}_2)}{\sqrt{\Lambda_1\Lambda_2}}\right\} \\ \qquad\quad -\dfrac{1}{8}m_1 m_2 B^{(1)}\Lambda_1\left(\dfrac{1}{\Lambda_1} + \dfrac{1}{\Lambda_2}\right)^2 |z_1|^2. \end{cases}$$

It remains to understand the topology of the level hypersurfaces of the restriction of Q_0 to the ellipsoid defined by the dominant term of the equation fixing the norm of the angular momentum (supposed to be close to its maximal value corresponding to circular horizontal motions), that is

$$|u_1|^2 + |u_2|^2 + \frac{\Lambda_1 + \Lambda_2}{\Lambda_2}|z_1|^2 = \epsilon^2.$$

If ϵ is small, the topology will be the same as for the reduced secular Hamiltonian itself. There are exactly three critical points, corresponding to the eigenplanes of Q_0:[32] two of them correspond to horizontal ellipses whose perihelia are in conjunction ($\Sigma^{\rightarrow}$) or opposition ($\Sigma^{\leftrightarrow}$); the third one $\Sigma^{\odot}$ ($u_1 = u_2 = 0$) corresponds to a pair of circles with non-zero inclination (see Figure 4). As singular points of the reduced secular Hamiltonian, $\Sigma^{\odot}$ and $\Sigma^{\leftrightarrow}$ are elliptic while $\Sigma^{\rightarrow}$ is hyperbolic. But as singularities of the Hamiltonian vector field in K^4, the three singular points are elliptic (i.e., the eigenvalues of their linearized part are imaginary).

Remark. For the planar problem, the averaged system is completely integrable. Indeed, it has two degrees of freedom and two commuting integrals, the energy and the angular momentum. For a global study and the application to the existence of various type of quasi-periodic solutions of the planar Three-Body Problem, see [Fe1].

---

[30] There are sign discrepancies with the formula that Poincaré attributes to Tisserand in Section 48; in particular, having different conventions for inclinations, his formula for mutual inclination is $i_1 - i_2$ instead of $i_1 + i_2$.

[31] These linearized equations are obviously first integrals of the Hamiltonian system defined by$Q$.

[32] Because of the symmetries, they are the same for the Hamiltonian and for its quadratic part Q_0.

4. Periodic solutions 1) Local existence by continuation

4.1. Pénétrer dans une place jusqu'ici réputée inabordable

The periodic solutions are the subject, in 1884, of the first work of Poincaré on the Three-Body Problem ([P4], announced by a note in the "Compte Rendus de l'Académie des Sciences" in 1883). They appear in Section 36 of Chapter III of *The New Methods.* As a good astronomer, Poincaré called a solution periodic if it becomes such in some rotating frame, as is, in the first approximation, the case of the motion of the Moon seen from the Earth. In other words, a solution $x(t)$ is periodic in this sense if there exists a *period* T such that for all t, $x(t+T) = \mathcal{R}x(t)$, where $\mathcal{R}$ is a rotation, that is if it is a periodic solution of the system reduced by the rotational symmetry.

These solutions play a key role in the whole work, and their study by continuation methods prefigures aspects of the theory of singularities:

> Le problème que nous allons traiter ici est le suivant : Supposons que, dans les équations (1),[33] les fonctions X_i dépendent d'un certain paramètre μ ; supposons que dans le cas de $\mu = 0$ on ait pu intégrer les équations, et qu'on ait reconnu ainsi l'existence d'un certain nombre de solutions périodiques. Dans quelles conditions aura-t-on le droit d'en conclure que les équations comportent encore des solutions périodiques pour les petites valeurs de μ ?[34]

Announcing the existence of an infinity of periodic solutions of the planar planetary Three-Body Problem when the planetary masses are sufficiently small, Poincaré justified his efforts as follows:

> Il semble d'abord que ce fait ne puisse être d'aucun intérêt pour la pratique. En effet, il y a une probabilité nulle pour que les conditions initiales du mouvement soient précisément celles qui correspondent à une solution périodique. Mais il peut arriver qu'elles en diffèrent très peu, et cela a lieu justement dans les cas où les méthodes anciennes ne sont plus applicables. On peut alors avec avantage prendre la solution périodique comme première approximation, comme *orbite intermédiaire*, pour employer le langage de M. Gyldén.
>
> Il y a même plus : voici un fait que je n'ai pu démontrer rigoureusement, mais qui me paraît pourtant très vraisemblable.
>
> Étant données des équations de la forme définie dans le numéro 13 et une solution particulière quelconque de ces équations, on peut toujours trouver une solution périodique (dont la période peut, il est vrai, être très longue), telle que la différence entre les deux solutions

[33] These are the equations $dx_i/dt = X_i(x_1, \dots, x_n)$, ; $i = 1, \dots, n$.

[34] The problem we are going to address is the following: suppose that, in equations (1), the functions X_i depend on some parameter μ; suppose that, when $\mu = 0$ one has integrated the equations and recognized the existence of some periodic solutions. Under what conditions shall we be able to conclude that for small values of μ the equations still possess periodic solutions?

> soit aussi petite qu'on le veut, pendant un temps aussi long qu'on le veut. D'ailleurs, ce qui nous rend ces solutions périodiques si précieuses, c'est qu'elles sont, pour ainsi dire, la seule brèche par où nous puissions essayer de pénétrer dans une place jusqu'ici réputée inabordable.[35]

Then, describing Hill's researches on the Moon, he further explains the importance of the construction of approximating periodic solutions:

> Supposons que, dans le mouvement d'un astre quelconque, il se présente une inégalité[36] très considérable. Il pourra se faire que le mouvement véritable de cet astre diffère fort peu de celui d'un astre idéal dont l'orbite correspondrait à une solution périodique.
>
> Il arrivera alors assez souvent que l'inégalité considérable dont nous venons de parler aura sensiblement le même coefficient pour l'astre réel et pour cet astre idéal; mais ce coefficient pourra se calculer beaucoup plus facilement pour l'astre idéal dont le mouvement est plus simple et l'orbite périodique.
>
> C'est à M. Hill que nous devons la première application de ce principe. Dans sa théorie de la Lune, il remplace ce satellite dans une première approximation par une Lune idéale, dont l'orbite est périodique. Le mouvement de cette Lune idéale est alors celui qui a été décrit au *n*° 41, où nous avons parlé de ce cas particulier des solutions périodiques de la première sorte, dont nous devons la connaissance à M. Hill.
>
> Il arrive alors que le mouvement de cette Lune idéale, comme celui de la Lune réelle, est affecté d'une inégalité considérable bien connue sous le nom de *variation* ; le coefficient est à peu près le même pour les deux Lunes. M. Hill calcule sa valeur pour sa Lune idéale avec un grand nombre de décimales. Il faudrait, pour passer au cas de la nature, corriger le coefficient ainsi obtenu en tenant compte des excentricités, de l'inclinaison et de la parallaxe. C'est ce que M. Hill eût sans doute fait s'il avait achevé la publication de son admirable Mémoire.[37]

---

[35]It seems at first that this fact cannot be of any interest in practice. Indeed, there is zero probability that the initial conditions of motion be precisely those which correspond to a periodic solution. But it may happen that they differ very little, and this occurs precisely where the old methods no longer apply. We can then use the periodic solution as a first approximation, as an *intermediate orbit*, to use the language of Mr. Gyldén. There is even more: here is a fact that I could not prove rigorously, but which nevertheless seems very likely to me. Given equations of the form defined in n° 13 and an arbitrary solution of these equations, one can always find a periodic solution (with a period which, admittedly, may be very long), such that the difference between the two solutions be arbitrarily small. In fact, what makes these solutions so precious to us, is that they are, so to say, the only opening through which we can try to enter a place which, up to now, was deemed inaccessible.

[36]that is, a deviation from the elliptic motion due to the action of the Sun.

[37]Suppose that, in the motion of some star, there is a very strong inequality. It may happen that the true motion of this star will differ very little from the motion of an ideal star whose orbit would correspond to a periodic solution. It will often happen that the large inequality which we

Moreover, we shall see that periodic solutions play a key role in the proof of non-integrability of the Three-Body Problem. The various objects attached to them are of common use in dynamics: the *characteristic exponents* are introduced in Chapter IV, the *asymptotic solutions*[38] in Chapter VII, the *doubly asymptotic solutions*[39] in Chapter XXXIII.

4.2. The three "sorts"

In Section 39, Poincaré recalls the classification of periodic solutions into three sorts[40] from his 1884 paper:

> ... j'ai été conduit à distinguer trois sortes de solutions périodiques: pour celles de la première sorte, les inclinaisons sont nulles et les excentricités très petites ; pour celles de la deuxième sorte, les inclinaisons sont nulles et les eccentricités finies ; enfin, pour celles de la troisième sorte, les inclinaisons ne sont plus nulles.[41]

These three sorts are continuations of periodic solutions which exist for the limit problem $F_0(x)$ which consists in two uncoupled Kepler Problems. Namely

1) two circular motions in the same plane;

2) two elliptic motions in the same plane with resonant frequencies, perihelia in conjunction or opposition and initial mean longitudes both equal to zero;

3) two circular motions in different planes with resonant frequencies.

In the first sort, the eccentricities of the planets are small and they have no inclination (planar problem); in the limit where the masses vanish, the orbits become circular, which is the singularity of the secular Hamiltonian: if n_1, n_2 are the frequencies of the two circular Keplerian motions, the limit solution has period $T = \frac{2\pi}{n_1 - n_2}$: indeed after time T the two bodies have turned respectively by the angles $\frac{2\pi n_1}{n_1 - n_2}$ and $\frac{2\pi n_2}{n_1 - n_2}$ whose difference is 2π. In terms of the coordinates

---

just mentioned will have approximately the same coefficient for the real star and for this ideal star; but this coefficient will be much more easily computed for the ideal star whose motion is simpler and the orbit is periodic. It is to Mr Hill that we owe the first application of this principle. In his Lunar theory, he replaces this satellite in the first approximation by an ideal Moon, whose orbit is periodic. The motion of this ideal Moon is then the one described in section n° 41, where we discussed of this special case of periodic solutions of the first sort, which we know thanks to Mr Hill. It then arrives that the motion of this ideal Moon is affected of a very large inequality, well known under the name of variation; the coefficient is approximately the same for the two Moons. Then Mr Hill computes its value for his ideal Moon with a large number of decimals. In order to treat the case of nature, one would need to correct the coefficient obtained in this way by taking into account the eccentricities, the inclination and the parallax. This is most probably what Mr Hill would have done if he had completed the publication of his admirable Memoir.

[38] Today called *stable or unstable manifolds*, or more generally *invariant manifolds*.

[39] Today called *homoclinic* or *heteroclinic* solutions.

[40] in French: les trois sortes.

[41] I was led to distinguish three sorts of periodic solutions: for those of the first sort, the inclinations are equal to zero and the eccentricities are very small, for those of the second sort, the inclinations are equal to zero and the eccentricities are finite; finally, for those of the third sort, the inclinations are no longer equal to zero.

introduced in 3.1, the existence proof for $\mu \neq 0$ amounts to writing that the expressions $\Lambda_1, \Lambda_2, u_1 e^{i\lambda_1}, u_2 e^{i\lambda_2}, \lambda_1 - \lambda_2$ take the same values at times 0 and T, which expresses that after time T the mutual distances between the bodies have regained their initial value[42]. Writing these as functions of the initial conditions and the period and taking into account the first integrals, Poincaré shows that the continuation of such solutions to small values of μ is possible provided n_1 is not an integer multiple of $n_2 - n_1$ (this ensures that the ad hoc Jacobian determinant does not vanish). He also notices that the periodic solutions obtained in this way have necessarily the following property: the bodies are in *symmetrical*[43] *conjunction.* at some time t_0, and in *symmetrical opposition* at time $t_0 + T/2$.

The most famous example of a periodic solution of the first sort (in the case of the Restricted Problem, that is when one of the masses is infinitesimal, see 9.1) is Hill's intermediate orbit used in his study of the Lunar problem. Describing Hill's result in Par. 41, Poincaré corrects a wrong guess of Hill about the global continuation of his solution (see 4.3); see the answer sent by Hill to Poincaré at the address `http://www.univ-nancy2.fr/poincare/chp/text/hill18920120.xml`.

In the second sort, the inclinations are still zero but the eccentricities are finite; in the limit one gets elliptic motions with the same direction of semi-major axes and conjunctions or oppositions at each half-period. In the third sort, eccentricities are small but inclinations are finite and the limit motions are circular but inclined. Here we are facing a *bifurcation* problem: indeed, we start from the degenerate and completely integrable situation where families of periodic solutions of the reduced problem which exist for $\mu = 0$ are expected to break for $\mu \neq 0$ and give rise to isolated periodic solutions. More precisely, we start from a *resonant torus* of dimension two, corresponding to motions along two Keplerian ellipses of the form $l_1(t) = n_1 t + l_1^0, l_2(t) = n_2 t + l_2^0$, where the values of L_1 and L_2 are chosen so that there exists $T \in \mathbb{R}$ and $k_1, k_2 \in \mathbb{N}$ with $n_1 T = k_1 2\pi, n_2 T = k_2 2\pi$.

In what he called "Examen d'un important cas d'exception"[44], Poincaré shows that, when as in the (reduced) planetary problem, the unperturbed Hamiltonian F_0 depends only on some of the action variables[45], that is when the phase variables are grouped into two subsets, the fast (or Keplerian) ones and the slow (or secular) ones, the continuation for $\mu \neq 0$ of the unperturbed periodic orbits by periodic orbits of the *same* period T, amounts to finding critical points, either

---

[42] This is because $ue^{i\lambda} = \sqrt{2(L-G)}e^{il}$; hence the given functions determine a_1, a_2, e_1, e_2 (that is the shape of both ellipses), l_1, l_2 (that is the position of the bodies on the ellipses) and $g_1 - g_2$ (that is the relative position of the ellipses).

[43] "Symmetrical" means that the velocities are orthogonal to the line joining the bodies; this implies a symmetry with respect to this line of the motions of the bodies at times symmetrical with respect to t_0

[44] Consideration of an important exceptional case.

[45] In this case the Hessian of F_0 with respect to the action variables vanishes identically, which forbids using the trick of replacing F_0 by a function $\varphi(F_0)$ of F_0 whose Hessian does not vanish; introduced by Poincaré this trick allows in some cases (see for instance 9.1) to regain the possibility of using directly the implicit function theorem.

non degenerate or of odd multiplicity (in particular maximum of minimum) of the average R of the perturbing function F_1 along the unperturbed periodic solutions.

In the planar problem, R depends on the 4 initial values H^0, l_1^0, l_2^0, h^0 of the reduced coordinates (see 2.4) (those of L_1, L_2 have already been fixed in order to fix the unperturbed frequencies). Moreover, one can fix the phase, that is the origin of time, by setting $l_1^0 = 0$, which leaves the three variables H^0, l_2^0, h^0. In the spatial case, it depends on the 6 initial values of the reduced coordinates which can be reduced to the five $G_1^0, G_2^0, l_2^0, g_1^0, g_2^0$.

At this point, Poincaré notices that, *provided one makes the ansatz that at some time* (*chosen to be* $t = 0$) *the bodies are in symmetric conjunction*, that is if in addition to l_1^0 we suppose that $l_2^0 = g_1^0 = g_2^0 = 0$, the dominant term of the expansion in eccentricities of R is nothing but the reduced secular Hamiltonian[46]. Hence he is able to relate the existence of periodic solutions of the second and third sort to the singularities of the reduced secular system. He then appeals to the computations of the quadratic part of the averaged Hamiltonian done by the "founders of Celestial Mechanics" as they are reproduced in Tisserand's treatise[47] [Ti] (see 3.3) to compute the critical points of the reduced Hamiltonian. This allows him in particular to show the existence of periodic solutions of the third sort different from those of the second sort, i.e., with non zero inclinations. In his characteristic style, he concludes:

> Cela ne veut pas dire qu'il n'existe pas également des solutions périodiques de la troisième sorte pour lesquelles il n'y ait pas de conjonction symétrique; il se pourrait en effet que, la fonction R admît d'autres maxima ou minima que ceux qui correspondent au cas de $l_0' = g_0 = g_0' = 0$. Il y aurait donc lieu de revenir sur cette question.[48]

Notice that this ansatz is crucial to make precise the link with the singularities of the secular system; indeed, because of the assumed resonance between the mean motions of the planets, the secular system is not a good approximation to the actual system.

Figure 4 is an attempt to represent the unperturbed periodic solutions from which the second and third sort periodic solutions bifurcate: the eight-dimensional reduced phase space of the spatial Three-Body Problem is fibered over the four-dimensional reduced secular space K^4, diffeomorphic to the complex projective

---

[46] More precisely, one must also avoid those mean motion resonances which give rise to periodic terms of order ≤ 2 in eccentricities, the first case being the (1,1)-resonance, that is $n_1 = n_2$, which contributes a periodic term whose coefficient is of order zero in eccentricities.

[47] Mentioning this work in the introduction of the *Lessons*, Poincaré writes (see 6.4):

> I had not to do again what he had done and done well.

[48] This does not mean that solutions of the third sort for which there is no symmetric conjunction do not exist; it could indeed be the case that the function R possesses other maxima or minima that the ones which correspond to the case of $l_0' = g_0 = g_0' = 0$. Hence one should come back to this question.

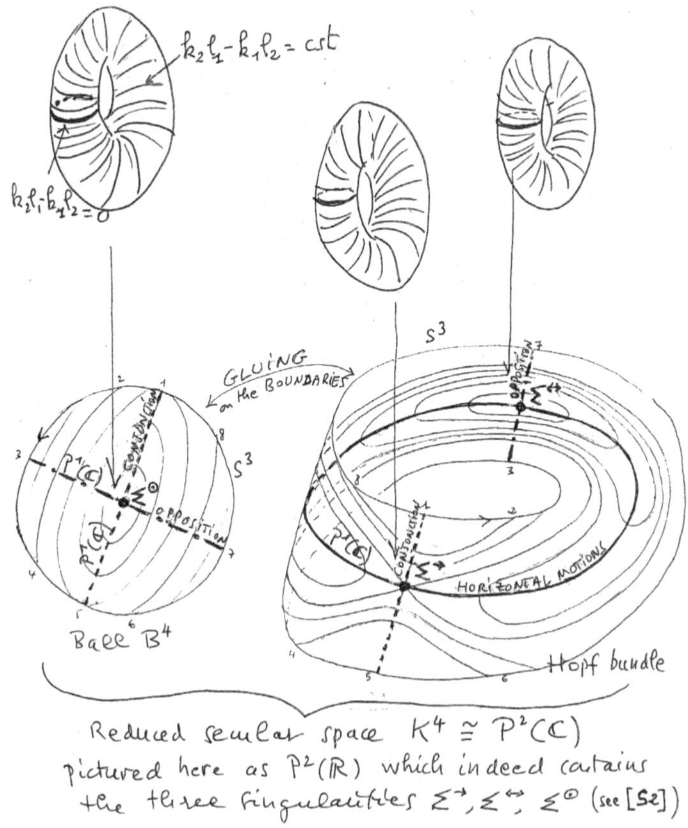

FIGURE 4. Unperturbed solutions at the origin of 2nd and 3rd sort orbits.

plane $P_2(\mathbb{C})$. In the representation, I have replaced K^4 by its "real part", diffeomorphic to the real projective plane $P_2(\mathbb{R})$, obtained by gluing along their boundary (diffeomorphic to a circle) a Moebius band and a disc. This is not completely stupid as, because of the symmetries of the secular Hamiltonian (invariance under complex conjugacy, see 3.3), the critical points we are interested in are contained in this real part. The fibers of the projection on K^4 are diffeomorphic to $\mathbb{T}^2 \times \mathbb{R}^2$. Once a resonant pair of values of the semi-major axes (more precisely a resonant pair (Λ_1, Λ_2)) has been fixed, the fibers become 2-tori, parametrized by the mean longitudes (λ_1, λ_2) and each of these tori is filled by a family of periodic solutions. The ansatz amounts to choosing one (in fact two, corresponding to opposition of conjunction) of these periodic solutions. Among those, the ones lying over the critical points of the reduced secular system will be ones for which Poincaré proves that they can be continued after the perturbation is turned on.

4.3. and some others

For example the ones in the neighborhood of an equilibrium position, mentioned in Sections 51 and 52. This is one the few times Poincaré mentions a relative equilibrium, namely the Euler ones L_1, L_2 of the Restricted Problem (see Figure 12), where the Moon would stay at constant distance of the Earth and always be in opposition or conjunction. He shows the existence in the neighborhood of these equilibria of periodic solutions (all this of course in the rotating frame, see 9) which agree with Hill's intuition even if Hill's reasoning (continuation of his solution 4.2) was wrong.

5. Quasi-periodic solutions 1) Formal aspects: Lindstedt series

> Il y a entre les géomètres et les astronomes une sorte de malentendu au sujet de la signification du mot *convergence*.[49]

The first chapter of the second volume of *The New Methods* is devoted to the definition of what Poincaré calls an *asymptotic series*. It opens with a clarification of the different conceptions geometers and astronomers have of the notion of convergence, the astronomers paying essentially attention to the way the first terms decrease or increase (least term summation). Well aware of the practical side of the question, he specifies:

> Les deux règles sont légitimes : la première, dans les recherches théoriques ; la seconde, dans les applications numériques. Toutes deux doivent régner, mais dans deux domaines séparés et dont il importe de bien connaître les frontières.
>
> Les astronomes ne les connaissent pas toujours d'une façon bien précise, mais ils les franchissent rarement ; l'approximation dont ils se contentent les maintient d'ordinaire beaucoup en deçà ; d'ailleurs leur instinct les guide et, s'il les trompait, le contrôle de l'observation les avertirait promptement de leur erreur.[50]

The excellent survey [Ra1] by J.P. Ramis renders unnecessary for me to further discuss Poincaré's conceptions on divergent series. I shall just mention the introduction of the *Lessons* [P12], where Poincaré studies the "ancient methods" without paying attention to the problems of convergence. Alluding to *The New Methods*, he says:

---

[49]There is between geometers and astronomers some sort of misunderstanding about the meaning of the word *convergence*.

[50]The two rules are legitimate: the first, in theoretical researches; the second, in the numerical applications. Both must reign, but within two distinct domains the boundaries of which it is important to know well. Astronomers do not always know them very precisely, but they seldom cross them; the approximation they are satisfied with usually maintains them much further; besides, their instinct guides them and, if it did cheat them, the control by the observation would quickly warn them of their error.

> D'autre part, dans ces Volumes, j'ai poussé l'approximation beaucoup plus loin que ne l'exige la pratique ; j'ai pu ainsi faire ressortir des circonstances tout à fait imprévues, dont l'importance analytique est très grande, mais qui n'ont aucun intérêt pour l'astronome praticien, et n'en acquerront que le jour où la précision des observations sera beaucoup plus grande qu'aujourd'hui, ou quand on voudra comparer des observations s'étendant sur une longue suite de siècles.[51]

If one is content with replacing centuries by hundreds of millions of years and observations by numerical simulations, J. Laskar's works (see [Las1, Las2]) attest the relevance of this sentence.

The search for quasi-periodic solutions of the equations, while a purely mathematical problem (behaviour of the solutions in infinite time!) has nevertheless astronomical significance if one is interested in the behaviour on very long periods of time.[52] It is intimately linked with the method of *averaging* already discussed in Section 3. Indeed, averaging or (*normal forms*) and secular systems will appear naturally in what follows.

Paraphrasing Poincaré, I must say that my way of exposing these chapters will differ quite a lot from Poincaré's exposition but that the series I shall obtain are the same as his. Also, I shall not discuss the attributions: Newcomb? Lindstedt? Gyldén? Delaunay? Bohlin? Let us say Poincaré for simplicity, if not completely accurate, this does not seem to be outrageously wrong. The interested reader may look in particular at [PC] and at the transparencies of a conference by P. Nabonnand (see [Na]) on the correspondence between Lindstedt and Poincaré.

5.1. What is a Lindstedt series?

The second volume of *The New Methods of Celestial Mechanics* is devoted to the study of the perturbation series, the main tool used by astronomers in order to "solve" the equations of motion. Immediately after a chapter dedicated to the notion of asymptotic expansion, for the study of which I refer to [Ra1], Poincaré shows the existence of quasi-periodic "solutions" of equations (1), of the form

$$\begin{cases} x = x^0 + \mu\Phi_1(w) + \mu^2\Phi_2(w) + \cdots, \\ y = w + \mu\Psi_1(w) + \mu^2\Psi_2(w) + \cdots, \end{cases} \qquad (L_\mu)$$

where the Φ_j and the Ψ_j are mappings from $\mathbb{T}^N$ to $\mathbb{R}^N$ and

$$w = (w_1, w_2, \ldots, w_N), \quad w_i(t) = \overline{w}_i(\mu) + n_i(\mu)t.$$

[51] On the other hand, in these Volumes, I pushed the approximation much further than needed in practice; this allowed me to put forward quite unforeseen circumstances, the analytical importance of which is very high, but which do not have any interest for the practitionning astronomer, and will become of interest only when the precision of observations will be much higher than today, or when one will be willing to compare observations taking place along a long sequence of centuries.

[52] Because of velocities close to the velocity of light are involved, this is particularly relevant in the study of the motion of particles in an accelerator, the relevant period of time becoming relatively short; see [M4].

Such solutions look like what a family depending on μ of solutions (2) of a family of integrable systems (1) defined by $F_0(\chi, \mu)$ (the notations are those of 2.5 with (x, y) replaced by (χ, w) and the parameter μ added) would become after a family of "coordinate changes" as shown in Figure 5.

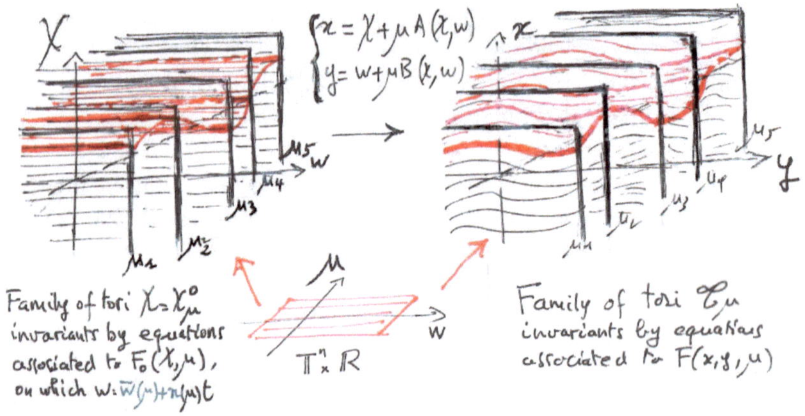

FIGURE 5. Families of invariant tori defined by Lindstedt series.

Using the implicit function theorem, one finds that the corresponding integral curves are confined for each μ to a torus $\mathcal{T}_\mu$ of the form

$$x = x^0 + \mu x^1(y) + \mu^2 x^2(y) + \cdots$$

invariant by the flow of the equations (1) defined by $F(x, y, \mu)$. In particular, they define motions in which the bodies stay forever in past and future at finite distance from each other without any collision ever happening. The grain of salt, which explains the brackets, is that the "solutions" and the "coordinate changes" are all given by formal series in μ, the convergence of all of which would contradict Poincaré's non-integrability result. The novelty is that these series are devoid of the disturbing *secular terms*, containing powers of the time t outside the sines and cosines, which, growing without limits, did spoil the series given by the "old methods" and appear now as mere artefacts of the Taylor expansion[53]. Poincaré

[53]It is fair to quote d'Alembert who, in 1759, writes somewhat bitterly in volume 9 of the *Encyclopédie* [En, C6]:

"Je ne dois pas oublier d'ajouter 1°. *que ma méthode pour déterminer le mouvement de l'apogée, est très-élégante & très-simple, n'ayant besoin d'aucune intégration, & ne demandant que la simple inspection des coefficients du second terme de l'équation différentielle;* 2°. *que j'ai démontré le premier par une méthode rigoureuse, ce que personne n'avoit encore fait, & n'a même fait jusqu'ici, que l'équation de l'orbite lunaire ne devoit point contenir d'arcs de cercle; si on ajoute à cela la maniere simple & facile dont je parviens à l'équation différentielle de l'orbite lunaire, sans avoir besoin pour cela, comme d'autres géomètres, de transformations & d'intégrations multipliées ; & le détail que j'ai donné ci-dessus de mes travaux & de ceux des*

calls these "solutions" *Lindstedt series* because the fist terms had been obtained by Lindstedt, but he is the first who proves their existence, as he himself explains:

> Mais il y a une autre difficulté plus grave; on constate aisément que la méthode est applicable dans les premières approximations, mais on peut se demander si l'on ne sera pas arrêté dans les approximations suivantes; M. Lindstedt n'avait pu l'établir rigoureusement et conservait même à ce sujet quelques doutes. Ces doutes n'étaient pas fondés et sa belle méthode est toujours légitime; je l'ai démontré d'abord par l'emploi des invariants intégraux dans le *Bulletin astronomique, t. III, p.* 57, puis, sans me servir de ces invariants, dans les *Comptes rendus, t. CVIII, p.* 21.[54]

5.2. Different types of Lindstedt series

In order to understand the problems of convergence of the Lindstedt series, one needs to understand the choice one has when defining them. Consider again the case when

$$F(x,\mu) = F_0(x) + \mu F_1(x) + \cdots$$

does not depend on the angles y. For any value of μ, the phase space $\mathbb{T}^n \times \mathbb{R}^n$ is foliated by invariant tori of the form $x = \text{cst}$. A Lindstedt series is obtained in this case by choosing for each μ one of these invariant tori, depending nicely on μ. This leads to three main types: series with variable frequencies, series with fixed frequencies, or series with fixed energy and fixed projective frequencies. All this is implicit in Poincaré who, nevertheless chooses as does Lindstedt, the series with variable frequencies because they are those which are the most opposed to those with fixed frequencies which are given by the old methods. We shall write

$$\omega(x,\mu) = \frac{\partial F}{\partial x}(x,\mu) = \omega^0(x) + \mu\omega^1(x)\cdots$$

---

autres géomètres, on conviendra, ce me semble, que j'ai eu plus de part à la théorie de la lune que certains mathématiciens n'avoient voulu le faire croire".

"I must not forget to add 1°. that my method for determining the motion of the lunar apogee, is very elegant and very simple, needing no integration, and requiring only the simple inspection of coefficients of the second term of the differential equation; 2°. that I was the first to prove by a rigorous method, something that no one had yet accomplished, & that has not been accomplished until now, that the lunar orbit equation need contain no arcs at all; if one adds to this the simple and easy way by which I obtain the differential equation of the Lunar orbit, without having, as other geometers, to use for that multiple transformations and integrations, and the detail given above of my works and of the works of the other geometers, I think one will agree that I have more contributed to the Lunar theory than some mathematicians had wanted people to believe."

[54]But there is another, more serious, difficulty; one checks easily that the method can be applied in the first approximations, but one can ask if one will not be stopped in the next approximations; M. Lindstedt had not been able to prove it and he was even somewhat doubting it. These doubts were not founded and his beautiful method is always legitimate; I first proved it in the *Bulletin astronomique*, t. III, p. 57, by using the integral invariants, then without using them in the *Comptes rendus*, t. CVIII, p. 21.

the *frequency vector*. If the *initial conditions* $(\bar{x}, \bar{y}) = (x^0, y^0)$ are fixed independently of μ, one obtains a family of quasi-periodic solutions of the family of equations (H_μ), whose frequencies $n(\mu) = \omega(x^0, \mu)$ depend in general of μ. It is the most trivial example of a Lindstedt series.

If on the contrary one wants the frequencies to be fixed, that is independent of the parameter μ, one must chose initial conditions $\big(\bar{x}(\mu), \bar{y}(\mu)\big)$ satisfying the equation

$$\omega\big(\bar{x}(\mu), \mu\big) = \omega\big(x^0, 0\big) = \omega^0(x^0) = n^0\,, \quad \bar{x}(0) = x^0\,.$$

The implicit function theorem grants the possibility of such a choice for small enough μ as long as the matrix

$$\frac{\partial \omega}{\partial x}\big(x^0, 0\big) = \frac{\partial \omega^0}{\partial x}\big(x^0\big) = \frac{\partial^2 F_0}{\partial x^2}\big(x^0, 0\big)$$

is invertible (one says that the Hamiltonian F_0 is *isochronically non degenerate* at the point x^0).

Finally, if one wants the *energy* to be independent of μ, one can only fix the frequencies up to a common scaling factor, i.e., projectively; it is enough to choose initial conditions satisfying

$$\begin{cases} \omega\big(\bar{x}(\mu), \mu\big) = \lambda(\mu) n^0, \\ F\big(\bar{x}(\mu), \mu)\big) = \text{constant.} \end{cases}$$

This is possible as soon as

$$\begin{pmatrix} \frac{\partial^2 F_0}{\partial x^2}(x^0) & \frac{\partial F_0}{\partial x}(x^0) \\ tr\,\frac{\partial F_0}{\partial x}(x^0) & 0 \end{pmatrix}$$

is invertible, that is as soon as, at the point x^0, the Hamiltonian F_0 is *isoenergetically non-degenerate.*

5.3. From the old methods to the new ones and back

Let us consider now a family $F_\mu(x, y) = F(x, y, \mu)$ of Hamiltonians which are still completely integrable but without supposing that (x, y) be action-angle coordinates for F_μ. That is, we suppose that they come from action-angle coordinates (χ, w) through the symplectic change of coordinates

$$x_i = \chi_i + \mu \cos w_i, \quad y_i = w_i, \quad i = 1, \ldots, n,$$

which transforms F into

$$C(\chi, w) = C_0(\chi) + \mu C_1(\chi) \text{ (with necessarily } C_0 = F_0),$$

whose solutions are quasi-periodic, of the form

$$\chi = \bar{\chi}, \quad w = \bar{w} + \varpi(\bar{\chi}, \mu)t = \bar{w} + \varpi^0(\bar{\chi})t + \mu\varpi^1(\bar{\chi})t.$$

One obtains in this way the solutions of the Hamiltonian system (H_μ) associated with F in the form of Lindstedt series of one of the three forms:

1) *Series with variable frequencies* $\varpi(x^0,\mu) = n^0 + \mu n^1$:

$$x = x^0 + \mu\cos\bigl(y^0 + n^0t + \mu n^1t\bigr), \quad y = y^0 + n^0t + \mu n^1t.$$

2) *Series with fixed frequencies* $n^0 = \varpi(x^0,0)$:

$$x = x^0 + \mu\bigl(\alpha_1 + \cos(y^0 + n^0t)\bigr) + \mu^2\alpha_2 + \cdots, \quad y = y^0 + n^0t.$$

3) *Series with fixed energy and fixed projective frequencies:*

$$x = x^0 + \mu\bigl[\alpha_1 + \cos\bigl(y^0 + \lambda(\mu)n^0t\bigr)\bigr] + \mu^2\alpha_2 + \cdots, \quad y = y^0 + \lambda(\mu)n^0t.$$

The use of the strange notations

$$\bar\chi(\mu) = x^0 + \mu\alpha_1 + \mu^2\alpha_2 + \cdots, \quad \bar w(\mu) = y^0$$

for the initial conditions, is motivated by the choice of staying as close as possible to Poincaré's notations.

All these series are of the form

$$x = \bar\chi(\mu)+\mu\cos\bigl(y^0+\varpi^0\bigl(\bar\chi(\mu)\bigr)t+\mu\varpi^1\bigl(\bar\chi(\mu)\bigr)t\bigr),\ y = y^0+\varpi^0\bigl(\bar\chi(\mu)\bigr)t+\mu\varpi^1\bigl(\bar\chi(\mu)\bigr)t.$$

If we expand them in series of powers of μ, one obtains "old-fashioned" series

$$\begin{cases} x =x^0 + \mu\bigl(\alpha_1 + \cos\bigl(y^0 + n^0t\bigr)\bigr) \\ \qquad + \mu^2\left[\alpha_1 - \left(\dfrac{\partial\varpi^0}{\partial x}(x^0)\cdot\alpha_1 + n^1\right)t\sin\bigl(y^0 + n^0t\bigr)\right] + O(\mu^3), \\ y =y^0 + n^0t + \mu\left(\dfrac{\partial\varpi^0}{\partial x}(x^0)\cdot\alpha_1 + n^1\right)t + O(\mu^2), \end{cases}$$

which contain *secular terms* and are developed into the fixed frequencies $n^0 = \varpi^0(x^0)$. Knowing that Lindstedt had originally tried to suppress these secular terms by letting the frequencies vary but maintaining fixed the initial conditions, one could think that it is the shift from fixed to variable frequencies which makes the difference between old and new methods but what we have just done shows that the same aim may be attained a priori via an infinite number of ways, and in particular by keeping expansions with fixed frequencies: it suffices to make the initial conditions depend on μ in a well-chosen way. For example, the first step must eliminate the term

$$\frac{\partial\varpi^0}{\partial x}(x^0)\cdot\alpha_1 + n^1,$$

which is the term of order one in μ of $\varpi\bigl(\bar\chi(\mu),\mu\bigr)$.

Remark. The fact that the mapping $(\chi,w)\mapsto(x,y)$ is symplectic is easily checked directly but it is better to notice that it admits the *generating function* (another invention of Poincaré) $S(\chi,y) = \chi\cdot y + \mu\sum\sin y_i$, which means that

$$x = \frac{\partial S}{\partial y}(\chi,y), \quad w = \frac{\partial S}{\partial \chi}((\chi,y), \quad \text{that is} \quad dS(\chi,y) = x\cdot dy + w\cdot d\chi,$$

from which follows the identity $0 = d^2S = dx\wedge dy - d\chi\wedge dw$.

5.4. One torus or a foliation by tori? (Poincaré's ambiguities)

In the above examples, each solution of F_μ is quasi-periodic of the form (L_μ). In the phase space $D \times \mathbb{R}^n$, the closure of the integral curve defined by a convergent Lindstedt series is, when μ is fixed, a torus $\mathcal{T}_\mu$ whose dimension lies between one (periodic orbit) if there exists a non zero real number T and integers p_i such that for all i one has $n_i(\mu)T = p_i$, and n (quasi-periodic orbit of general type) if no *resonance* exists between the frequencies $n_i(\mu)$, that is if there is no relation of the form

$$k \cdot n(\mu) = \sum_{i=1}^{n} k_i n_i(\mu) = 0\,,$$

in which $k = (k_1, k_2, \ldots, k_n)$ are integers among which at least one is different from zero. Therefore, we are naturally led to look, as Poincaré did, not for formal particular solutions of equations (H_μ), but for *formally invariant tori* of these equations. Moreover, the structure of the solutions (L_μ) shows that, at the formal level, these tori should admit angular coordinates $\theta = (\theta_1, \ldots, \theta_n)$ such that the restriction of the equations becomes a constant vector field: in the non-resonant case where the closure of the integral curve (L_μ) has dimension n, such coordinates θ are obviously $(w_1, w_2, \ldots, w_n)$ which satisfy

$$\frac{dw_i}{dt} = n_i(\mu), \quad i = 1, \ldots, n.$$

Together with the density in the torus of the integral curves of the flow, this last property is a major constraint on an invariant torus of a Hamiltonian system: indeed, we notice after Michael Herman that the form Ω induced on such a torus by the symplectic form

$$dx \wedge dy = \sum_{i=1}^{n} dx_i \wedge dy_i,$$

which is invariant under the flow of equations (H_μ), is necessarily constant in the θ coordinates:

$$\Omega = \sum_{i=1}^{n}\sum_{j=1}^{n} c_{ij}\, d\theta_i \wedge d\theta_j.$$

But, the standard symplectic form $dx \wedge dy$ being the coboundary of the *Liouville form* $x{\cdot}dy = \sum_{i=1}^{n} x_i dy_i$, Ω itself is a coboundary and, being constant, it cannot but vanish identically. Finally, in the non-resonant case, we are looking for a formally invariant n-torus on which the form induced by the symplectic form vanishes. Today, such a torus is said to be *Lagrangian.* The very structure of (L_μ) shows that these tori must be looked for close to the tori $\{x^0\} \times \mathbb{T}^n$, that is as the graphs of a formal mapping Σ_μ from $\mathbb{T}^n$ to D. Being Lagrangian is then equivalent to the existence of a mapping $y \mapsto S_\mu(y)$ whose derivative is Σ_μ:

$$\Sigma_\mu(y) = \frac{dS_\mu}{dy}(y).$$

The classical *Hamilton–Jacobi theory* tells us that the only condition which S_μ must satisfy is to be a global solution of the *Hamilton–Jacobi equation*

$$F\left(\frac{dS_\mu}{dy}(y), y, \mu\right) = c(\mu),$$

where $c(\mu)$ is a constant (recall that morally, μ is fixed). This is a simple translation of the fact that such invariant Lagrangian submanifolds are nothing but geometric solutions of the Hamilton–Jacobi equation. Unfortunately in the absence of other hypotheses, a Lagrangian torus $\mathcal{T}_\mu$ which is the graph of the derivative Σ_μ of a global solution S_μ of the Hamilton–Jacobi equation, has no reason to possess coordinates such that the Hamilton equations (H_μ) become a constant vector field (i.e., a quasi-periodic one). This last property is only ensured in case the torus belongs to an n-parameter family $\chi = (\chi_1, \chi_2, \ldots, \chi_n)$ of such tori, defined by a *complete solution* $S_\mu(\chi, y)$ of the Hamilton–Jacobi equation:

$$F\left(\frac{\partial S_\mu}{\partial y}(\chi, y), y, \mu\right) = C(\chi, \mu). \qquad (HJ_\mu)$$

This is because such a complete solution is the *generating function* of a symplectic diffeomorphism $\mathcal{D}_\mu$

$$(\chi, w) \mapsto (x, y),$$

implicitly defined by the identity

$$dS_\mu = w \cdot d\chi + x \cdot dy,$$

and in the coordinates (χ, w) the Hamiltonian $F(x, y, \mu)$ takes the form $C(\chi, \mu)$ independent of the angles w, hence is *completely integrable*: Our torus being defined by $\chi = \bar{\chi}$ and parametrized by the angles w, the equations of motion read

$$\frac{dw}{dt} = \frac{\partial C}{\partial \chi}(\bar{\chi}, \mu).$$

Of course, the only thing which we used is the existence of the family for χ in an infinitesimal neighborhood of $\bar{\chi}$: it is in fact enough to be able to take a derivative with respect to χ at the point $\bar{\chi}$. This remark will play a fundamental role in the next section.

On these points, Poincaré is slightly ambiguous: looking for a non-resonant torus, he makes it depend on parameters with respect to which he takes derivatives in order to find the frequencies. He does not care, at least explicitly, about the possibility of encountering denominators equal to zero in the coefficients of the series which define the generating function[55] $S_\mu(\chi, y)$. Nevertheless, one can make sense of all this thanks to Émile Borel who taught us how to take the derivative of a function defined on a Cantor set.

[55] This is because, contrarily to what we do in the next section, he works with formal series whose coefficients themselves are formal series coming from the multiplication of formal series.

5.5. Existence of the Lindstedt series in the non-degenerate case

Warning! My presentation, though close to Poincaré's, differs in that I immediately address the individual convergence of the coefficients of the series, while in Chapter X, Poincaré treats them also as formal series,[56] waiting Chapter XIII for all convergence questions.
We are looking for a "function" $S_\mu(\chi, y) = S(\chi, y, \mu)$ of the form

$$S(\chi, y, \mu) = \chi \cdot y + \sum_{i=1}^{\infty} \mu^i S_i(\chi, y)$$

whose partial derivatives $\frac{\partial S_i}{\partial \chi}$ et $\frac{\partial S_i}{\partial y}$ are 2π-periodic in the variables y_j, and which satisfies formally an equation of the form

$$F\left(\frac{\partial S}{\partial y}(\chi, y, \mu), y, \mu\right) = C(\chi, \mu) = C_0(\chi) + \sum_{i=1}^{\infty} \mu^i C_i(\chi). \qquad (HJ_\mu)$$

The choice as terms of order zero terms in μ of $\chi \cdot y$, that is of a generating function of the Identity, is dictated by the fact that, as $F(x, y, 0)$ is independent of the angles y, there is no need to modify this first term. Note that the word "function" must be taken with a grain of salt as we deal with formal expressions for which no domain of definition is given. Our goal is indeed to give all this a meaning.

The periodicity constraints imply that, up to an arbitrary function of χ (corresponding to a harmless angular phase shift), the function S_i may be written

$$S_i(\chi, y) = \alpha_i \cdot y + s_i(\chi, y),$$

where $\alpha_i = (\alpha_{i,1}, \dots, \alpha_{i,n})$ belongs to $\mathbb{R}^n$, and s_i is 2π-periodic with mean value zero in the variables y_j.
The identification of powers of μ in the series expansion of (HJ_μ) gives then

$$\omega^0(\chi) \cdot \frac{\partial S_1}{\partial y}(\chi, y) + F_1(\chi, y) = C_1(\chi), \qquad (E_1(\chi))$$

where we have used the notation

$$\omega^0(x) = \frac{\partial F_0}{\partial x}(x)$$

for the frequency vector of the unperturbed Hamiltonian $F_0(x)$.
The parameter $\chi = x^0$ being fixed, the determination of a particular invariant torus (depending on μ) is possible as long as x^0 is *sufficiently non resonant*, that is as long as it belongs to one of the subsets $D_{K,\nu}$ of D defined as follows:

$$D_{K,\nu} = \left\{x^0 \in D\,,\, \forall k \in \mathbb{Z}^n - \{0\}\,,\, |k \cdot \omega^0(x^0)| \geq \frac{K}{|k|^\nu}\right\}.$$

[56] which he multiplies as formal series with formal series coefficients ...

The Fourier series of $s_1(x^0, y)$,

$$s_1(x^0, y) = \sum_{k=(k_1,\dots,k_n)\in\mathbb{Z}^n-\{0\}} \sigma_k(x^0)e^{ik\cdot y},$$

is uniquely determined from that of $F_1(x^0, y)$,

$$F_1(x^0, y) = \sum_{k=(k_1,\dots,k_n)\in\mathbb{Z}^n} f_k(x^0)e^{ik\cdot y},$$

by the identities

$$\sigma_k(x^0) = f_k(x^0)/(ik \cdot \omega^0(x^0)) .$$

It is not difficult to see that, as the coefficients of the Fourier series of an analytic function decrease quicker than any power of $|k|$ and as the increase of the *small denominators* $ik \cdot \omega^0(x^0)$ is polynomial in $|k|$, the series so obtained defines an analytic function of y as long as x^0 belongs to one the the $D_{K,\nu}$.
Concerning α_1 and $C_1(x^0)$, they must be chosen in such a way as to satisfy

$$C_1(x^0) - \omega^0(x^0) \cdot \alpha_1 = \frac{1}{(2\pi)^n}\int_{\mathbb{T}^n} F_1(x^0, y)\, dy,$$

where the right-hand side is the *averaged system*, analogue in the non-degenerate case of the one studied in Section 3.2.
One resolves in the same way, by induction, the equations which come from the identification of the terms in $\mu^j, j \geq 2$ in the series expansion of (HJ_μ): the identities are of the form

$$\omega^0(\chi) \cdot (\partial S_j/\partial y)(\chi, y) + \Phi_j(\chi, y) = C_j(\chi), \qquad (E_j(\chi))$$

where the term $\Phi_j(\chi, y)$ is a polynomial in the $\frac{\partial S_i}{\partial y}(\chi, y), i < j$, whose coefficients are analytic in χ, y.
Once chosen the α_j, one obtains, for any x^0 belonging to one of the $D_{K,\nu}$, a formal series in μ, $S(x^0, y, \mu)$, whose all coefficients are analytic functions of y, and a formal series in μ, $C(x^0, \mu)$. The problem is now to "differentiate" these series with respect to $\chi_1, \dots, \chi_n$.
Taking formally the derivative of the equation $(E_j(\chi))$ with respect to the parameter χ_k, one obtains new equations

$$\omega^0(\chi) \cdot \frac{\partial^2 S_j}{\partial \chi_k \partial y}(\chi, y) + \frac{\partial \omega^0}{\partial x}(\chi) \cdot \frac{\partial S_j}{\partial y}(\chi, y) + \frac{\partial \Phi_j}{\partial \chi_k}(\chi, y) = \frac{\partial C_j}{\partial \chi_k}(\chi),$$

of the same form as above. Solving inductively these equations and the ones corresponding to higher derivatives, one determines, for each element x^0 of $D_{K,\nu}$, $S_j(\chi, y)$ and the $C_j(\chi)$ as formal series[57] in $(\chi - x^0)$ (that is in the $(\chi_i - x_i^0), i = 1, \dots, n$) whose coefficients are analytical in y, which satisfy the equations $(E_j(\chi))$.

[57] Notice that, for $j = 1$, the two formal series in $(\chi - x^0)$,

$$C_1(\chi) - \omega^0(\chi) \cdot \alpha_1 \quad \text{and} \quad \frac{1}{(2\pi)^n}\int_{\mathbb{T}^n} F_1(\chi, y)\, dy$$

are identical. In particular, $C_1(\chi)$ is defined and analytic for all χ.

As we already noticed, this is enough [58] to ensure the quasi-periodicity of the motion on the formal Lagrangian torus defined by the equation $x = \frac{\partial S}{\partial y}(x^0, y, \mu)$. One deduces the existence of Lindstedt series, implicitly defined by

$$\begin{cases} x = x^0 + \sum_{i=1}^{\infty} \mu^i \alpha_i + \sum_{i=1}^{\infty} \mu^i \frac{\partial s_i}{\partial y}(x^0, y), \\ y + \sum_{i=1}^{\infty} \mu^i \frac{s_i}{\chi}(x^0, y) = w = \bar{w}(\mu) + \frac{\partial C}{\partial \chi}(x^0, \mu)t. \end{cases}$$

If, as Poincaré does, one chooses to equal all the α_i to zero, one obtains Lindstedt series with variable frequencies. If, on the contrary, one chooses the α_i in such a way[59] that $\frac{\partial C_i}{\partial \chi}(x^0)$ vanishes for all $i \geq 1$, one obtains Lindstedt series with fixed frequencies.

Remarks. 1) In [Pos], Pöschel shows that the $S_i(\chi, y)$ (resp. the $C_i(\chi)$) are C^∞ *functions in the sense of Whitney* on each closed subset $D_{K,\nu} \times T^n$ (resp. $D_{K,\nu}$). This is equivalent to saying that they may be extended (in a non unique way) to C^∞ functions on the whole of $D \times \mathbb{T}^n$ (resp. D). The first result of this type, for mappings, is due to V. Lazutkin [Laz].

2) At the end of Section 147, Poincaré notices that the construction may be done without appealing to infinite sums at intermediary stages by truncating the Fourier series far enough to be entitled to putting the rest in the small perturbation:

> Voici donc ce qu'on sera conduit à faire : dans la série F_1, tous les termes, sauf un nombre fini d'entre eux, pourront être regardés comme du même ordre de grandeur que μ ; mais il y en aura qui seront du même ordre de grandeur que μ^2, d'autres, plus petits encore, qui seront du même ordre de grandeur que μ^3, etc. Dans les autres séries F_2, F_3, on trouvera de même des termes de ces divers ordres de grandeur.
>
> Nous pourrons donc écrire en général
>
> $$F_i = F_{i,0} + F_{i,1} + F_{i,2} + \cdots + F_{i,k}, \ldots,$$
>
> $F_{i,k}$ représentant ceux des termes de F_i qui peuvent être regardés comme du même ordre de grandeur que μ^k. Ces termes sont en nombre fini. Cette manière de décomposer F_i comporte évidemment un assez grand degré d'arbitraire.[60]

This trick, which has the inconvenient of losing trace of the analyticity in the parameter, was used by Arnold in [A1, A3].

[58] Obviously, order one is already enough.

[59] This is possible as soon as the unperturbed Hamiltonian F_0 is isochronically non-degenerate (see 5.2).

[60] The translation is an exercise for the reader.

5.6. Coping with the degeneracies ...

At the end of Section 128, Poincaré speculates that it is the degeneracy of the Three-Body Problem which prevented Lindstedt from discovering the possibility of writing down series with fixed frequencies. And indeed such series cannot exist because, as the precessions of perihelia and nodes vanish with the perturbation,[61] the slow frequencies must depend on the parameter μ, i.e., one can only expect Lindstedt series whose frequencies are of the form $\big(\omega^0(x'^0), \mu\,\omega^1(x^0)\big)$, where $x^0 = (x'^0, x''^0)$ is the decomposition of the action into fast and slow.

In order to show that Lindstedt series still exist in this case, Poincaré starts Chapter X by noticing that their existence in the non degenerate case applies to the averaged system in the neighborhood of circular and horizontal motions of the two planets:

> On peut faire des principes du Chapitre précédent une application importante à l'étude de certaines équations que les astronomes ont souvent considérées.[62]

Indeed, after replacing the symplectic polar coordinates $H_1, H_2, Z_1, Z_2, h_1, h_2, \zeta_1, \zeta_2$ defined in 2.2 and 2.3, by analogue coordinates which transform the quadratic part of the secular Hamiltonian into a sum of squares, and taking as a small parameter ϵ the distance to the origin (that is to zero eccentricities and inclinations), one can find a formal (in ϵ) change of coordinates which eliminates the angles at any order *provided the coefficients of the squares admit no relation with integer coefficients.* Poincaré comments:

> Ainsi on peut satisfaire formellement aux équations qui définissent les variations séculaires par des séries trigonométriques de la forme de MM. Newcomb et Lindstedt ... Ce résultat aurait été envisagé par Laplace ou Lagrange comme établissant complètement la stabilité du système solaire. Nous sommes plus difficiles aujourd'hui parce que la convergence des développements n'est pas démontrée; le résultat n'en est pas moins important.[63]

This is nothing but putting the secular system into *Birkhoff normal form* (see [AKN]), that is finding a formal[64] change of coordinates transforming the Poincaré coordinates $u_j = \xi_j + i\eta_j, z_j = p_j + iq_j$ into coordinates

$$\tau_j + i\sigma_j = \sqrt{2\rho_j}e^{i\omega_j}, j = 1, \ldots, 4,$$

[61] what Arnold calls a *proper degeneracy.*

[62] One can make of the principles of the former chapter an important application to the study of some equations that the astronomers have often considered.

[63] Hence one can formally satisfy the equations defining the secular variations by trigonometrical series similar to those of MM. Newcomb and Lindstedt ... This result would have been considered by Laplace or Lagrange as completely establishing the stability of the solar system. We are more demanding nowadays, because the convergence of these expansions is not proved; the result is nevertheless important.

[64] For the planar Three-Body Problem, the secular system can be reduced to a 1-degree of freedom system and hence is completely integrable; in this case, the procedure is not only formal.

such that the secular Hamiltonian depends only of their moduli and not on their arguments:

$$H_1 = \sum_{j=1}^{4} A_j(\tau_j^2 + \sigma_j^2) + \sum_{1 \leq j,k,}^{4} B_{j,k}(\tau_j^2 + \sigma_j^2)(\tau_k^2 + \sigma_k^2) + \cdots$$

The first step is transforming the quadratic part of the averaged system into a sum of squares by an orthogonal change of coordinates. Remembering the special form of the quadratic part (see 3.3) one sees that one can at this level make two independent linear changes of coordinates, one with the horizontal part (perihelia, eccentricities) to which correspond the *horizontal frequencies* A_1 and A_2, and the other with the vertical part (nodes and inclinations) (to which correspond the *vertical frequencies* A_3 and A_4).

Once the formal Birkhoff normal form is built (supposing there is no resonances between the Aj), one eliminates the fast angles by a formal change of coordinates similar to the one used in the non degenerate case. This is possible because *replacing F_0 by $F_0 + \mu\langle F_1\rangle$ (where $\langle F_1\rangle$ is the result of averaging the fast angles in F_1) has removed the degeneracy.* One could also have started by eliminating the fast angles independently of the form of the secular part, a process to which is attached today the name of Von Zeipel. In all cases, one has to stay outside a neighborhood of the set of values of the fast actions (here the semi-major axes) for which there is a mean motion resonance, in which case some combinations of the fast angles become slow and cannot be eliminated by averaging; this is the domain of *Bohlin series* (see 7), where resonant tori break into lower-dimensional tori and their asymptotic manifolds.

We have in fact disregarded the difficulty of very small eccentricities and inclinations: indeed, the singular point of the secular system which corresponds to a pair of horizontal and circular ellipses gives rise in the full system to two-dimensional invariant tori which after reduction become periodic solutions of the first sort (see 4.2). In order to be able to write Lindstedt series defining formal invariant Lagrangian tori very close to these (i.e., corresponding to quasi-periodic solutions whose eccentricities and inclinations are very small) it is necessary to center the symplectic polar coordinates one uses on these two-dimensional tori. This is exactly what Poincaré does in his Chapter XII.

Finally, if $x^0 = (x'^0, x''^0)$ belongs to a suitable subset of the actions, one obtains Lindstedt series whose frequencies are of the form

$$(\omega^0(x'^0), \mu\, \omega^1(x^0)).$$

This is as close as possible to the ones with fixed frequencies obtained in the non degenerate case.

5.7. ... but forgetting a resonance!

The method described in the former section works perfectly for the Three-Body Problem in the plane because the averaged system is in fact completely integrable: after reduction by the symmetry of rotation, that is fixing the angular momentum

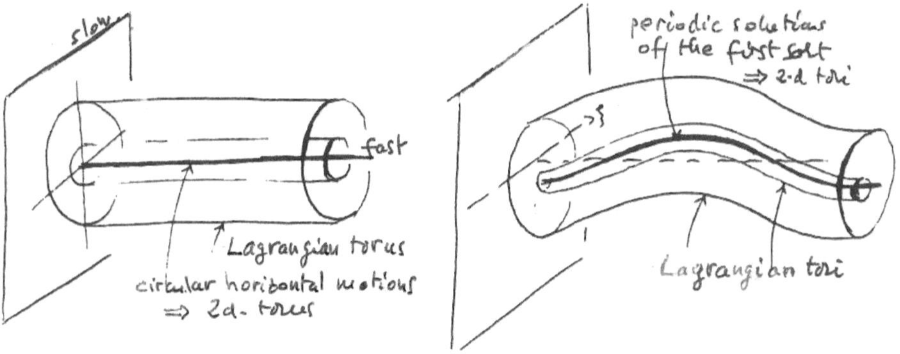

FIGURE 6. Invariant tori of very small eccentricities and inclinations. Reproduced from [S2] with permission of © IMCCE 1990.

and quotienting by the rotation group, one finds a system with only one degree of freedom, hence integrable. But a problem arises for the spatial problem: describing, at the end of Section 132, the application to the Three-Body Problem of the formal integration of the secular system (what we have called "putting the system into Birkhoff normal form"), Poincaré is well aware that, in order to prove the existence of the formal change of variables he must know that, writing, as we have just done, the quadratic part of the averaged system as a sum of squares $\sum_{j=1}^{4} A_j(\tau_j^2+\sigma_j^2)$,

> La seule hypothèse que nous devions faire, c'est qu'il n'y ait pas entre les quatre constantes A_1, A_2, A_3, A_4 de relation linéaire à coefficients entiers. La probabilité pour que cette relation existe est nulle, mais on peut encore se demander s'il n'y a pas une relation simple de cette forme qui soit assez près d'être satisfaite pour que les séries ne convergent plus que très lentement. On sait que Le Verrier a discuté cette question, mais il a dû la laisser indécise en ce qui concerne les planètes inférieures, parce que les masses en sont mal connues et que les coefficients A dépendent de ces masses.
>
> Il est clair que tout ce qui précède s'applique, sans qu'on ait rien à y changer, au cas où l'on aurait plus de trois corps.[65]

I must confess that I do not know of which series Poincaré says that they would "converge" very slowly, as a priori the averaged system is not integrable in the spatial case, but in any case, the phenomenon to which he assigned zero probability

[65]The only hypothesis we had to make is that, between the four constants A_1, A_2, A_3, A_4, there be no linear relation with integer coefficients. The probability for such a relation to exist is equal to zero, but one can still ask whether there is no simple relation of this form which would be almost satisfied so that the series would converge very slowly. One knows that Le Verrier discussed this question but he had to leave it unsettled for the inner planets because the masses are not well known and the coefficients A depend on these masses. It is clear that all of the above applies, without having to change anything, in case one would have more than three bodies.

does indeed occur *for the spatial problem* as one has *always* two resonances between the secular frequencies A_i:

1) the invariance under rotation forces one of the vertical frequencies, say A_4 to be equal to zero. This does not happen in the planar case because circles are invariant under a rotation of their plane. But in $\mathbb{R}^3$, the plane containing a pair of coplanar circles can still be rotated non trivially, which accounts for a symplectic eigenplane with 0 frequency.

2) whatever be the masses, one has

$$A_1 + A_2 + A_3 + A_4 = 0.$$

The first resonance is harmless. Indeed, it would disappear after reduction but there is more: the reduction is unnecessary because one can show that, precisely because of the symmetries of the problem, the terms in the Birkhoff normal form which would be affected by a denominator equal to zero are not present.
The second one is also harmless but in a subtler way. Discovered in full generality (i.e., for the spatial planetary $(1+n)$-body problem for any n) by Michael Herman, and known today under the name of *Herman's resonance*, it was implicitly known in the lunar theory since the beginning of the Newtonian theory through the fact that, if one takes into account only the terms of the first-order in the planetary masses, the theory of perturbation gives practically the same value of 18 years for the period of the direct motion of the lunar apogee and the retrograde motion of its line of nodes (see [AA] where it is pointed out that Poincaré mentions this resonance in [P7]).
Well aware of the first resonance (but somewhat forgetting it), Poincaré overlooked the second one, which makes his proof of the existence of Lindstedt series in the spatial case incomplete[66] but, fortunately, the Herman resonance disappears after full reduction of the rotation symmetry, and this allows to restore the proof. More precisely (see [MRL]), in the Three-Body Problem, it spoils the Birkhoff normal form only from terms of order at least ten in the eccentricities and inclinations and, even more, it does not appear at all in the full formal Birkhoff normal form after the partial reduction which consists in fixing only the direction of the angular momentum (we shall come back to this in Section 11). Finally, it has no incidence on the dynamics and moreover, it disappears if one extends the averaging at the second order of the masses.[67] A striking illustration of this last fact is the difference between the value of 18 years for the period of the lunar apogee and the actual value of approximately 9 years.[68] This discrepancy had casted a major doubt on the Newtonian theory until Clairaut explained in 1750 that taking into account some terms of higher order [in the lunar problem, the small parameter μ is the

[66] It is this same mistake, overlooking the Herman resonance, that Arnold will make in 1963, see Section 11 ... the Oulipo would say that Poincaré had made a mistake by anticipation.

[67] In some sense, sticking to the frequencies at the secular singularity to build expansions is too rigid, in a way as rigid as were the old methods making Taylor expansions at fixed frequencies and hence introducing secular terms.

[68] On the contrary, the value of 18 years for the period of the node is close to the actual one.

ratio of mean motions of the Moon and the Sun] did restore the correct value of 9 years.[69]

6. Periodic solutions 2) The source of complexity

We have asserted in Section 2.6 that the breaking of periodic invariant tori of the unperturbed system was the main cause of the extreme complexity of most of the solutions of the Three-Body Problem. This complexity is displayed on the one hand in the non-integrability and the divergence of Lindstedt series, both originating from the breaking of periodic invariant tori of the integrable approximations which gives rise to periodic solutions with *non-zero exponents*, on the other hand in the divergence of Bohlin's series which signs the *splitting of separatrices.* In the popular account [P6], also reproduced in [Ra1], Poincaré gives a very lucid account of the way he understands the role of periodic solutions:

> Les équations différentielles du problème des trois corps admettent un certain nombre d'intégrales qui sont connues depuis longtemps ; ce sont celles du mouvement du centre de gravité, celles des aires, celles des forces vives. Il était extrêmement probable qu'elles ne pouvaient avoir d'autres intégrales algébriques ; ce n'est cependant que dans les dernières années que M. Bruns a pu le démontrer rigoureusement. Mais on peut aller plus loin ; en dehors des intégrales connues, le problème des trois corps n'admet aucune intégrale analytique et uniforme ; les propriétés des solutions périodiques et asymptotiques, étudiées avec attention, suffisent pour l'établir. On peut en conclure que les divers développements proposés jusqu'ici sont divergents ; car leur convergence entraînerait l'existence d'une intégrale uniforme.[70]

6.1. Variational equations: exponents, stability, asymptotic solutions

Initiating the study of the stability of periodic orbits, Poincaré introduces their *characteristic exponents* and their *stable and unstable manifolds.* The famous mistake in his 1889 prize memoir [P1], where he had thought he had proved stability in the Restricted Problem, is about the intersection of these manifolds (see [BG, Y]). We shall come back to this topic in Section 9.

Chapter IV begins with the definition of the variational equations. Given a particular solution $x(t) = (x_1(t), \dots, x_n(t))$ of a differential equation $\frac{dx}{dt} = X(x)$ defined

[69] For a full account of this important episode, see [d'Al] and the recension [C6].

[70] The differential equations of the Three-Body Problem possess a number of integrals which have long been familiar; these are those of the motion of the center of mass, the area integrals, the energy. It was extremely unlikely that they could have other algebraic integrals; it is, however only in the recent years that Mr. Bruns has proved this rigorously. But we can go further; apart from the known integrals, the Three-Body Problem admits no analytic and uniform integral; a careful study of the properties of periodic and asymptotic solutions is enough to establish this. It can be concluded that the various developments proposed so far are divergent; for their convergence would imply the existence of a uniform integral.

in some domain of $\mathbb{R}^n$, the variational equation along φ is obtained by writing that $x(t)+\xi$ is also a solution and ignoring the terms in $\xi=(\xi_1,\ldots,\xi_n)$ of order two or more:

$$\frac{d\xi_i}{dt}=\frac{\partial X_i}{\partial x_1}\big(x(t)\big)\xi_1+\cdots+\frac{\partial X_i}{\partial x_n}\big(x(t)\big)\xi_n.$$

In order to stress the importance of these equations, Poincaré recalls that Hill used them in his "admirable theory of the Moon": starting with his periodic solution (see 4.1) which already gave a reasonable value of the inequality called variation, the study of the variational equations allowed him to compute the motions of the perigee and the node, and the coefficient of the *eviction.*

Having defined the exponents α_j of a linear differential equations with periodic coefficients via Floquet's theory[71], he turns his attention to the variational equation along a periodic solution of an autonomous differential equation: in this case, the exponents are the logarithms of the eigenvalues of the *monodromy* $d\varphi_T(x(t))$ at any point of the periodic solution; one of them is always equal to zero and moreover, to each independent first integral corresponds a zero eigenvalue and to each continuous symmetry corresponds a zero co-eigenvalue.

In the particular case of the (Hamiltonian) equations of dynamics

$$\frac{dx_i}{dt}=-\frac{\partial F}{\partial y_i},\quad \frac{dy_i}{dt}=\frac{\partial F}{\partial x_i},$$

he notices that, writing (ξ,η) and (ξ',η') two solutions of the variational equations,[72]

$$\sum_{i=1}^{n}\frac{d}{dt}\left[\eta_i'\xi_i-\xi_i'\eta_i\right]=0,$$

and analogous determinantal formulae for 4, 6, …, $2n$ solutions which are nothing but *the preservation by the flow of the symplectic form and its powers.*[73]

---

[71]Given a linear differential equation $\frac{dx}{dt}=A(t)x$ where the matrix $A(t)$ is periodic of period T and a *fundamental matrix solution* $X(t)$, that is an invertible matrix whose columns are a basis of solutions of the equation, one notices that the columns of $X(t+T)$ are also solutions, hence that there exists a *monodromy matrix* $\mathcal{M}$ such that $X(t+T)=X(t)\mathcal{M}$. The set of exponents is the spectrum, deprived of the obvious eigenvalue 1, of the monodromy matrix. It is well defined because a change of fundamental matrix solution leads to a conjugate monodromy matrix. Having recalled the theory, Poincaré cannot help adding:

> Je n'insiste pas sur tous ces points de détail. Ces résultats sont bien connus par les travaux de MM. Floquet, Callandreau, Bruns, Stieltjes, et, si j'ai donné ici la démonstration *in extenso* pour le cas général, c'est que son extrême simplicité me permettait de le faire en quelques mots.
>
> I do not insist on this minor point. These results are well known by the works of MM. Floquet, Callandreau, Bruns, Stieltjes and if if I have given here the proof *in extenso* for the general case, it is because its extreme simplicity allowed me to do it in a few words.

[72]ξ and η being respectively the small increases of x and y.

[73]For more on this, see Section 8.1.

Then, recalling that, if a linear transformation preserves a quadratic form, its characteristic polynomial is "reciprocal", he deduces[74] that the characteristic exponents are pairwise equal (in absolute value) and of opposite sign.[75]

The next step is to study the solutions which start close to the given periodic solution by writing power series for the solutions $\xi(t)$ of the (non linearized) equation $\frac{d}{dt}(x(t)+\xi) = X(x(t)+\xi)$. In Chapter VII, the last chapter of the second volume of *The New Methods*, Poincaré shows the existence and analyticity with respect to a parameter of what we call today *stable and unstable manifolds* of a periodic solution, respectively associated with the set of exponents with negative or positive real part (the so-called *hyperbolic part* of the spectrum). He does so by writing down convergent series expansions for the families of solutions they contain. He does not discard the *central* part corresponding to exponents whose real part is zero but is aware that convergence could fail. We shall come back to these invariant manifolds in the particular case of the Restricted Problem in Section 9.

Now comes the important fact heralded in Section 2.6: for the unperturbed Hamiltonian $F_0(x)$ of the general problem of dynamics, the variational equation along any periodic solution is trivial and hence the exponents are all equal to zero. Showing that after adding the small perturbation $\mu F_1(x,y)$ periodic solutions with non trivial exponents will appear is at the root of the non-integrability of the problem. Indeed, to the families of periodic solutions of the equations of dynamics, which depend analytically on the small parameter μ, are attached families of asymptotic solutions but, as all the exponents vanish when $\mu = 0$, the exponents and the solutions of the variational equations have series expansions not in powers of μ but in powers of its square root $\sqrt{\mu}$. Poincaré then deduces (after some work!) that the asymptotic solutions may be expressed as series expansions in powers $\sqrt{\mu}$, with coefficients given by series in powers of variables $w_j = A_j e^{\alpha_i t}$ (where the α_j are the characteristic exponents of the periodic solution), e^{it} and e^{-it}; he shows that these series diverge[76] but that, while divergent, they are *asymptotic series* in the sense he will formalize in Chapter VIII, which opens the second volume:

> On peut donc dire que les séries que nous avons obtenues dans le n° 108 représentent les solutions asymptotiques pour les petites valeurs de μ

---

[74]The proof is contested by A. Wintner in 1931 but it seems that Wintner forgot that the differential equation with periodic coefficients which is considered is the variational equation along a periodic solution of an *autonomous* equation.

[75]This theorem is rightly named by Kozlov "Poincaré–Lyapunov theorem". Indeed, in his book [Lya], published in 1892 by the mathematical society of Kharkov, Lyapunov proves this theorem and he explains that the result was known to him before the publication of [P1], and that he had communicated it in February 1890 to the Kharkov mathematical society.

[76]Another proof, based on the divergence of the Bohlin series, will be given at the very end of volume II.

> de la même manière que la série de Stirling représente les fonctions eulériennes.[77]

The intuitive idea behind these expansions is explained on the following Figure 7 in the case of 2 degrees of freedom: a continuous family of periodic solutions of the unperturbed problem gives rise to isolated *elliptic* and *hyperbolic* periodic solutions and hence, in the first approximation, to penduli of width $\sqrt{\mu}$ (*opening of a resonance*).

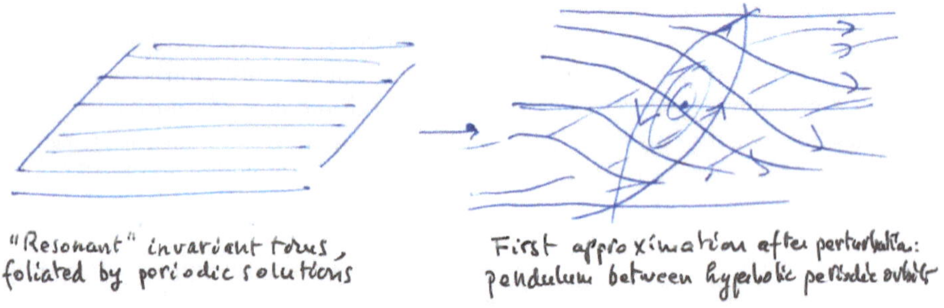

FIGURE 7. Opening of a resonance.

This is nothing but enriching Figure 3 with invariant manifolds: analytically, it corresponds to the family of integrable Hamiltonians

$$F_0(x) + \mu F_1(x,y) = \frac{1}{2}(x_1^2 + \cdots + x_n^2) - \mu\omega^2 \cos y_1.$$

We shall come back to this situation when studying Bohlin series in Section 7

6.2. Non-existence of uniform first integrals

Bruns had shown[78] the non-existence of first integrals of the Newtonian Three-Body Problem algebraic in the velocities, which do not arise from the ones which

[77] Hence one may say that the series that we have obtained in n° 108 represent the asymptotic solutions for the small values of μ in the same way as the Stirling series represent the Eulerian functions.

[78] Actually, Brun's paper contained a mistake which was noticed and corrected by Poincaré in [P9]. The conclusion of this paper is one more characteristic example of Poincaré's style:

> Le résultat de M. Bruns se trouve confirmé; je suis heureux d'avoir pu compléter son élégante analyse sur un point de détail.
>
> Hence the result of Mr. Bruns is confirmed. I am pleased to have been able to complete his elegant analysis on a small technical point.

It has nevertheless some savour to compare this modest ending with the corresponding statement in Poincaré's analysis of his own works:

> On sait que Bruns a démontré que le problème des trois corps ne saurait admettre d'autre intégrale algébrique que les intégrales classiques. Malheureusement dans sa démonstration subsistait une lacune grave et particulièrement délicate à combler. J'ai été assez heureux pour mettre la belle et ingénieuse démonstration de M. Bruns à l'abri de toute objection.

are consequence of the symmetries of the problem: energy and angular momentum. Using a completely different method, intimately related to the behaviour of periodic solutions, Poincaré shows in Chapters V and VI the non-existence of new first integrals analytic in x, y and μ, that is which are analytic in the (sufficiently small) masses of the planets. Comparing his result with Brun's theorem, he writes at the end of Section 85:

> Le théorème qui précède est plus général en un sens que celui de M. Bruns, ... Mais, en un autre sens, le théorème de M. Bruns est plus général que le mien ; j'établis seulement, en effet, qu'il ne peut pas exister d'intégrale algébrique pour toutes les valeurs suffisamment petites des masses ; et M. Bruns démontre qu'il n'en existe pour aucun système de valeurs des masses.[79]

If "generic" periodic solutions (i.e., solutions with non trivial exponents) were dense in the phase space, or even only if they formed a uniqueness set,[80] this would imply analytic non-integrability. This is because the existence of a complete set of almost everywhere independent commuting integrals[81] implies that every periodic orbit on which the integrals are independent has all its exponents equal to zero. Unfortunately, even if it is likely to be true, this property is not proved, hence the necessity of more analytic methods of proving non-integrability.

> One knows that Bruns has proved that the Three-Body Problem could not admit other algebraic integral than the classical ones. Unfortunately a serious gap remained in his proof which was very difficult to fill. I was fortunate enough to put the beautiful and clever proof of Mr. Bruns away from any objection.

For a complete proof of Brun's theorem, see [JT]; for a generalization (not really checked), see [Pa].

[79] The theorem above is in some sense more general than the one of M. Bruns, ... But, in another sense, the theorem of M. Bruns is more general than mine; indeed I prove only that there does not exist algebraic integrals valid for all sufficiently small values of the masses; and M. Bruns proves that such integral exists for no system of values of the masses whatsoever.

[80] i.e., a set such that an analytic function which is equal to zero on the set is identically zero.

[81] that is integrals $G, H, K, \ldots$, whose Hamiltonian flows all commute and hence define a possibly singular foliation of the phase space by invariant Lagrangian tori; this is equivalent (see [A2]) to the vanishing of all their *Poisson brackets* $\{G, H\} = \sum \frac{\partial G}{\partial x_j}\frac{\partial H}{\partial y_j} - \sum \frac{\partial H}{\partial x_j}\frac{\partial G}{\partial y_j}$, {G,K},{H,K},...

The actual proof[82] rests on a delicate analysis of the abundance of non-zero coefficients in the Fourier expansion in the angles y for the perturbing function $\mu F_1 + \cdots$, which amounts essentially to proving the "generic" behaviour of periodic solutions, that is the fact that their exponents are non-zero, and forbids them to form continua filling "periodic" invariant tori as is the case when $\mu = 0$ (see again 2.6).

In order to show how resonances do appear in the reasoning, let us examine the *non-degenerate* case, where we suppose that the Hessian of the unperturbed Hamiltonian $F_0(x)$ does not vanish, that is when the frequency map $x \mapsto \frac{\partial F_0}{\partial x}(x)$ is a diffeomorphism. Supposing that there is an integral $\Phi = \Phi_0 + \mu\Phi_1 + \cdots$, Poincaré develops the identity of commutation[83]

$$\{F, \Phi\} = \{F_0, \Phi_0\} + \mu\left(\{F_0, \Phi_1\} + \{F_1, \Phi_0\}\right) + \cdots = 0.$$

The first relation implies that Φ_0 does not depend on the angles y. If the Fourier expansions of F_1 and Φ_1 are respectively

$$F_1(x, y) = \sum_{m \in \mathbb{Z}^n} B_m(x) e^{im \cdot y}, \quad \Phi_1(x, y) = \sum_{m \in \mathbb{Z}^n} C_m(x) e^{im \cdot y},$$

the second relation means that the Fourier coefficients of Φ_1 must satisfy

$$\forall m, \quad C_m \sum m_j \frac{\partial F_0}{\partial x_j} - B_m \sum m_j \frac{\partial \Phi_0}{\partial x_j} = 0,$$

that is: $\sum m_j \frac{\partial \Phi_0}{\partial x_j}(x) = 0$ each time an action belongs to the *secular set*[84]

$$\mathcal{S} = \left\{x, \ \exists m = (m_1, \ldots, m_n); \sum m_j \frac{\partial F_0}{\partial x_j}(x) = 0 \quad \text{and} \quad B_m(x) \neq 0\right\}.$$

Notice that, the equation above being homogeneous in m, the relevant feature of the perturbing function is not whether an individual Fourier coefficients B_m is different from 0 but whether a *class* contains a non zero coefficient, where $B_{m'}$ and $B_{m''}$ are in the same class if $m'_1/m''_1 = m'_2/m''_2 = \cdots = m'_n/m''_n$.

---

[82] Comparing, in Section 85, this proof to the one he had given In the Memoir (see 6.5), Poincaré writes:

> Dans mon Mémoire des *Acta Mathematica* (t. XIII), je me suis servi pour établir le même point de l'existence des solutions périodiques et du fait que les exposants caractéristiques ne sont pas nuls. La démonstration que je donne ici ne diffère de celle des *Acta* que par la forme, mais elle se prête mieux à la généralisation qui va suivre.
>
> In my Memoir of the Acta Mathematica (t. XIII), in order to establish the same point, I used the existence of periodic solutions and the fact that the characteristic exponents are non zero. The proof I give here differs from the one in the Acta only by the form, but it is more apt to the following generalization.

[83] Recall that $\{F, G\} = \sum \frac{\partial F}{\partial x_j} \frac{\partial G}{\partial y_j} - \sum \frac{\partial G}{\partial x_j} \frac{\partial F}{\partial y_j}$ is the *Poisson bracket*.

[84] Poincaré says that a Fourier coefficient B_m of the perturbing function *becomes secular* if the actions x are such that $\sum m_j \frac{\partial F_0}{\partial x_j}(x) = 0$.

Figure 8 illustrates the definition of a class in the case of two degrees of freedom: the level curves of F_0 and Φ_0 must be tangent at each point of the secular set. If this set is *big enough*, for example if it is dense or if it is a uniqueness set, this will imply that the level sets of F_0 and Φ_0 coincide, and hence that Φ_0 is a function of F_0. In this case, after subtracting Φ_0 and dividing by μ one has a new integral of the form $\Phi_1 + \mu\Phi_2 + \cdots$ and the rest of the proof is similar.

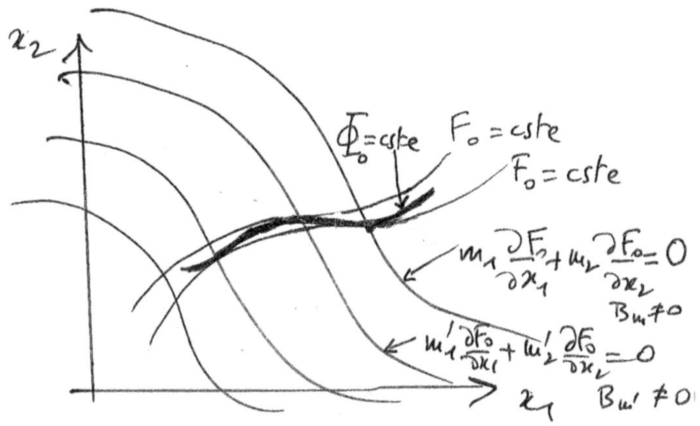

FIGURE 8. Being tangent along the secular set.

When there are more than two degrees of freedom, the proportionality of the gradients of F_0 and Φ_0 at x will now take place on the *Poincaré set*

$$\mathcal{P} = \Big\{ x, \exists m^{(1)}, \ldots, m^{(n-1)} \in \mathbb{Z}^n \text{ independent};$$
$$\forall k, \sum_j m_j^{(k)} \frac{\partial F_0}{\partial x_j}(x) = 0,\ B_{m^{(k)}}(x) \neq 0 \Big\}.$$

These values x of the actions define a *completely resonant* (i.e., filled with periodic solutions) torus of the unperturbed system and the fact that the $B_{m^{(k)}}$ are non zero, means that such a torus will be broken by the perturbation. Good references for this, with many examples, are [Koz, AKN].

As usual, this simple theory applies to the case of the Restricted Three-Body Problem, the one studied in the Memoir,[85] but not to the general Three-Body Problem because of the degeneracy of the Kepler Problem. The relevant Fourier expansions of F_1 and Φ_1 being $\sum B_{m_1,m_2} e^{i(m_1 l_1 + m_2 l_2)}$ and $\sum C_{m_1,m_2} e^{i(m_1 l_1 + m_2 l_2)}$,

[85] See Section 9; the Hamiltonian $F_0(L,G) = -\frac{K}{L^2} + G$ is degenerate (its Hessian vanishes) but Poincaré's trick of replacing it by $\exp F_0$ eliminates the degeneracy without modifying the integrability properties. Moreover, it follows from the form, computed by Leverrier (see [Cha]) of he Fourier expansion of the perturbing function, that the secular set is dense, hence the proof of the non-integrability.

where l_1, l_2 are the two mean anomalies (the fast angles) and B_m, C_m depend on the fast actions and all the slow variables, the slow variables introduce new terms in the relations between B_m and C_m and the Poincaré set must be defined in a subtler way. This is done in Chapter V, while the long Chapter VI ,"*Développement de la fonction perturbatrice*", is devoted to showing that the Fourier coefficients of the expansion in the mean anomalies of the perturbing function[86] satisfy the conditions established in Chapter V: associating to a Fourier expansion the function of complex variables obtained by replacing complex exponentials by complex variables, Poincaré generalizes to functions of two complex variables a method by Darboux, first used by Flamme, which relates the behaviour of the high rank Fourier coefficients of a periodic function $\sum_{n\in\mathbb{Z}} B_n e^{in\varphi}$ to the singularities of the complex function $\sum_{n\in\mathbb{Z}} B_n z^n$. After 66 pages of virtuous use of deformations of integration contours where he stresses the role of the coalescence of singular points of the integrand, he may conclude:

La non-existence des intégrales uniformes se trouve ainsi démontrée.[87]

6.3. Divergence of Lindstedt series

It is in Chapter XIII that Poincaré proves the divergence of the series with varying frequencies. It is essentially a consequence of the proof of the non-integrability result which asserts the breaking of the periodic invariant tori of the unperturbed system: if the Lindstedt series with varying frequencies did converge, this would imply the existence of such periodic tori, hence a contradiction.

Even if he considers it very unlikely, he is aware of the possibility of the convergence of Lindstedt series with fixed frequencies "satisfying some arithmetic conditions"; this is the famous Paragraph 149, which I quote in Section 11, putting it in the light of Arnold's "theorem" on the stability of planetary systems.

Remark. In his proofs of non-integrability and hence of the divergence of Lindstedt series, Poincaré does not make use of the subtler phenomena of separatrix splitting (see the question at the end of Section 10.1). A proof of analytic non-integrability based on the existence of such transversal homoclinic intersections was done by Jürgen Moser in an example due to Sitnikov, with two degrees of freedom [M5].

6.4. Back to the old methods: the "Lessons of Celestial Mechanics"

As already mentioned in Section 5, Poincaré was well aware of the different aims of geometers and astronomers; he knew perfectly that the divergence of the Lindstedt series did not diminish their practical interest and that the old methods were not to be rejected readily. In 1898, he had published in the "Bulletin Astronomique" [P11] a long article explaining how to recover the classical series expansions of the astronomers from the "new ones", insisting in particular on the fact that, according to the purpose being sought (a very precise representation of the position of the celestial bodies during a short time or a less precise representation but during a

[86] In particular the *principal part*, that is the inverse distance between the two planets (see Section 2.3).

[87] The non-existence of uniform integrals is thus rigorously proved.

very long time) the dominant terms in the series are not the same. A testimony of this attitude is the courses given in the Sorbonne and published under the name of *Lessons of celestial mechanics* [P12] between 1905 and 1910. Much less known[88] than the *New Methods*, this other great book on Celestial Mechanics is the redaction by people in the audience of courses at the Sorbonne; directly aimed at students, it examines the classical methods of perturbation. In the introduction, Poincaré warns his readers:

> Je n'emprunte aux méthodes nouvelles que leurs résultats essentiels, ceux qui sont susceptibles d'une application immédiate, en m'efforçant de les rattacher le plus intimement possible à la méthode classique de la variation des constantes.
>
> D'un autre côté, Tisserand s'est constamment préoccupé de reproduire aussi fidèlement qu'il a pu la pensée des fondateurs de la Mécanique céleste et, en effet, son Livre nous la rend tout entière sous une forme condensée. Je n'avais pas à refaire ce qu'il avait fait et bien fait.[89]

As *The New Methods*, the *Lessons* have three volumes: the first one, *General theory of planetary perturbations* (1905), starts with a thorough study of the general properties of the Three-Body Problem and, in particular of the various systems of coordinates convenient to study the planetary problem; this is followed by a remarkable exposition of the Keplerian elliptic motion, with in particular a complete proof of the analyticity[90] of what we have called the *Poincaré coordinates* (see 2.2). The chapters are: Lagrange's method; The Restricted Problem; Theory of secular perturbations; Poisson theorem; Symmetries in the developments, periodic solutions; Delaunay's method. The second volume is divided into two parts: 1) *Développement of the perturbing function* (1907), 2) *Theory of the Moon* (1909), while the third one is devoted to the *Theory of tides* (1910).

6.5. Complexifying Poincaré's method

The main new tool in order to prove non-integrability results was introduced by Ziglin in [Z]: starting with some particular solution of the equations, the obstruction to the existence of first integrals meromorphic in a neighborhood of this solution is found in the monodromy group of the variational equation along the complexification of the given solution. As Alain Albouy likes to say, Ziglin's method can be viewed as a complexification of the method originally used by Poincaré in

---

[88]in particular to the author, which explains the brevity of this section.

[89]To the new methods, I borrow only the essential results, those capable of immediate application, while striving to relate them as closely as possible to the classical method of the variation of constants. On the other hand, Tisserand paid constant attention to reproducing as faithfully as possible the thought of the founders of Celestial mechanics and, indeed, his book restitutes it in a compact form. I had no need to do again what he had done and done well.

[90]An amusing story: it happened to me to be the referee of a paper showing the analyticity of *Poincaré's coordinates* for the Kepler Problem. The result was less strong than Poincaré's and I had simply to refer to the first volume of the Lessons.

his Memoir to prove non-integrability: any first integral was shown to be dependent of the Hamiltonian along any "generic" (i.e., with non trivial exponents) periodic solution. The proof is extremely simple: Let $F(x,y)$ be a Hamiltonian, φ_t the flow of the corresponding Hamiltonian vector field[91] X_F and Φ a first integral of these equations[92]. If the solution $\varphi_t(z_0)$ is T-periodic, i.e., if $\varphi_T(z_0) = z_0$, one deduces from the constancy of F and Φ and the chain rule on the one hand, differentiating with respect to t the identity $\varphi_T \circ \varphi_t(z_0) = \varphi_t(z_0)$ on the other hand, that the monodromy matrix $d\varphi_T(z_0)$ of the periodic solution $\varphi_T(z_0)$ has 2 co-eigenvectors $dF(z_0)$ and $d\Phi(z_0)$ with co-eigenvalue 1 and one eigenvector $X_F(z_0)$ with eigenvalue 1:

$$dF(z_0)d\varphi_T(z_0) = dF(z_0),\ d\Phi(z_0)d\varphi_T(z_0) = d\Phi(z_0),\ d\varphi_T(z_0)X_F(z_0) = X_F(z_0).$$

Moreover $dF(z_0)X_F(z_0) = d\Phi(z_0)X_F(z_0) = 0$, from which one deduces that either this periodic solution has at least three exponents equal to zero, and hence is not generic, or the derivatives $dF(z_0)$ and $d\Phi(z_0)$ are dependent. After complexification, the variational equation becomes a linear differential equation along a Riemann surface and Ziglin's criterion of non-integrability involves the non commutation of some elements of the *monodromy group* of this equation which leads to the branching of solutions. A great advantage of the complexification is that the periodic solution is no more required to be generic. A *differential Galois group* version of this theorem was developed by Morales and Ramis [MR], and extended to variational equations of higher order by Morales–Ramis–Simó [MRS]. The criterion is the non-abelianity of the Lie algebra of the differential Galois group. Specific applications of these methods to the three (or more)-body problem were given by D. Boucher, J.J. Morales, A. Tsygvintsev, C. Simó, S. Simon, T. Combot, J.A. Weil, ... References will be found in [Au] and the more recent [Com1]. See also [Com2] which helps justifying the use of the Morales–Ramis theorem in most of the previous papers.

7. Resonances 1) Bohlin series

Resonances play a major role in celestial mechanics; the earliest mathematical studies seem to be those of Laplace on the 4:2:1 resonance between the Jovian satellites Ganymede, Europa and Io, and the 5:2 resonance between Jupiter and Saturn.

The neighborhood of a single resonance is, in the first approximation, well described by the equations of the product of simple rotators by a simple pendulum;[93] As written by Tisserand [Ti]:

---

[91] If $z = (x,y)$, $X_F(z) = \left(-\frac{\partial F}{\partial y}(x,y), \frac{\partial F}{\partial x}(x,y)\right)$ and $\frac{d\varphi_t}{dt}(z) = X\left(\varphi_t(z)\right)$. Recall that the monodromy is $d\varphi_T(z_0)$.

[92] The situation is of course symmetric and amounts to the vanishing of the Poisson bracket $\{F,\Phi\}$.

[93] Too many physics students ignore the phase portrait of the simple pendulum!

> If a dynamical system is in a state of resonance, then there exists a linear combination of the angle-variables, the so-called critical argument, that behaves like the angular coordinate of the simple pendulum. Depending on the initial conditions, it either librates about the point of stable equilibrium or it circulates around the clock.

The fundamental fact is that *this approximate system is completely integrable*, which is no more true in case of multiple resonance (see [GSV]).

7.1. Circulation, libration, separatrix: what is a Bohlin series?

> Reprenons les hypothèses et les notations du n° 125. Nous avons vu que dans l'application de la méthode du n° 125 il s'introduisait des diviseurs de la forme
> $$n_1^0 m_1 + n_2^0 m_2 + \cdots + n_n^0 m_n,$$
> les m_i étant entiers.[94]
>
> Il en résulte que cette méthode devient illusoire quand l'un de ces diviseurs devient très petit.[95]

This quotation opens the long Chapter XIX, devoted to simple resonances. The simplest illustration of the phenomena which are studied is the birth of a simple pendulum from a linear rotator:

$$F(x_1, \dots, x_n, y_1, \dots, y_n) = \frac{1}{2}(x_1^2 + x_2^2 + \cdots + x_n^2) + \mu \cos y_1 \quad (\mu \geq 0).$$

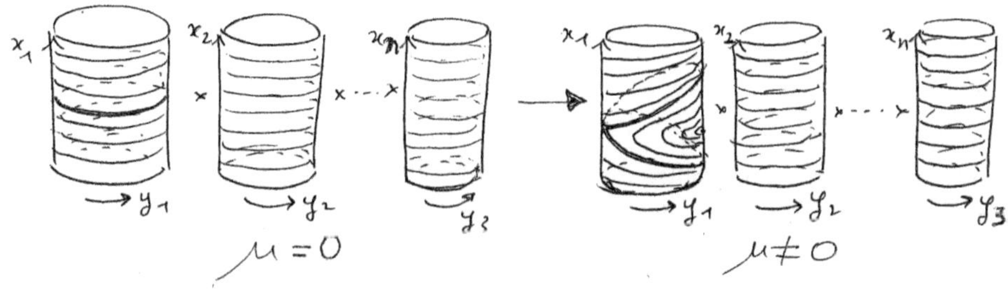

FIGURE 9. Opening of a simple resonance.

This is the case where $x^0 = (0, x_2^0, \dots, x_n^0)$ and $(m_1, m_2, \dots, m_n) = (1, 0, \dots, 0)$. In other words, the *resonant angle* is[96] $\sum m_i y_i = y_1$. Note that the actions $x_2, \dots, x_n$

[94] I remind you that with the notations of the *general problem of dynamics* (*see Section* 2.5) and my sign conventions, $n_i^0 = \frac{\partial F_0}{\partial x}(x^0)$ is the frequency of the invariant torus of the perturbed system with action $x = x^0$.

[95] Let us come back to the hypotheses and notations of n° 125. We have seen that, in the application of the method of n° 125, small denominators came up, of the form $n_1^0 m_1 + n_2^0 m_2 + \cdots + n_n^0 m_n$, where the m_i are integers. It results that this method become illusive when one of these divisors becomes very small.

[96] This choice of the resonant angle is not a restriction as one can reduce the general case to this one by a linear symplectic change of coordinates.

are integrals of the motion. We shall fix $x^0 = (0, x_2^0, \dots, x_n^0)$ and write

$$c_0 = F_0(x^0) = \frac{1}{2}((x_2^0)^2 + \cdots + (x_n^0)^2).$$

In this integrable model, three types of solutions coexist, depending on the value C of the energy:

the *circulation* type, if $C - c_0 > \mu$;
the *libration* type, if $-\mu < C - c_0 < \mu$,
the *separatrix* type if $C - c_0 = \mu$.

In the circulation range, the corresponding invariant tori are Lagrangian graphs and hence solutions of the Hamilton–Jacobi equation (compare with Section 5.4)

$$F\left(\frac{\partial S}{\partial y_1}(C, y), \frac{\partial S}{\partial y_2}(C, y), \dots, \frac{\partial S}{\partial y_n}(C, y), y_1, y_2, \dots, y_n, \mu\right) = C.$$

The solutions

$$S(C, y) = x_2^0 y_2 + \cdots + x_n^0 y_n \pm \int \sqrt{2(C - c_0 - \mu \cos y_1)}\, dy_1$$

admit an expansion in series of powers of μ (i.e., Lindstedt series) of the form $S = \pm\sqrt{2C} \sum_k \varphi_k(y_1)\,(\mu/C)^k$ but, as Poincaré notices, this expansion becomes useless in practice if $C - c_0$ is too close to μ and the expansion in powers of $\sqrt{\mu}$ has to be preferred.

In the libration range, the same holds true provided we first change variables in a way which turns the corresponding invariant tori into graphs.

For the separatrix type, which corresponds to any action $x^0 = (0, x_2^0, \dots, x_n^0)$ at which $n_1^0 = \frac{\partial F_0}{\partial y_1}(x^0)$ vanishes, the situation is different. Indeed, the solution of the Hamilton–Jacobi equation satisfies

$$\begin{aligned} S(c_0 + \mu, y) &= x_2^0 y_2 + \cdots + x_n^0 y_n \pm \int \sqrt{2\mu(1 - \cos y_1)}\, dy_1 \\ &= x_2^0 y_2 + \cdots + x_n^0 y_n \pm 2\sqrt{\mu} \int \sin \frac{y_1}{2}\, dy_1 \\ &= x_2^0 y_2 + \cdots + x_n^0 y_n \pm 4\sqrt{\mu} \cos \frac{y_1}{2}. \end{aligned}$$

Notice that S is no more 2π-periodic but only 4π-periodic in y_1. It is the image of an immersed torus which covers twice the circulation tori (see Figure 9). The two solutions are deduced from each other by a translation of 2π.

7.2. "Bohlin method" in the non-degenerate case

What Poincaré calls the Bohlin method, a conceptual improvement of the method used by Delaunay in his study of the motion of the Moon, was discovered in 1888 by K. Bohlin [Boh] and Poincaré independently (see the introduction of the memoir; in *The New Methods*, Poincaré acknowledges a slight anteriority of Bohlin). Supposing that F_0 depends on all the action variables[97] and that

$$n_1^0 = \frac{\partial F_0}{\partial x_1}(x^0) = 0$$

is the sole linear relation between the frequencies,[98] it consists, in each of the three cases, in searching for solutions of the above Hamilton–Jacobi equation[99] which are formal series of the form

$$S = S_0 + S_1\sqrt{\mu} + S_2\mu + S_3\mu\sqrt{\mu} + \cdots, \quad C = C_0 + C_2\mu + C_4\mu^2 + \cdots$$

with

$$\forall i, \ \frac{\partial S_0}{\partial y_i} = x_i^0 \quad \text{and} \quad \frac{\partial S_p}{\partial y_i} \ \text{periodic.}$$

Referring to [S3] for an analysis of some of the obscurities in Poincaré's exposition which I shall not discuss in depth, I shall be content with showing how the three types of solutions appear in the equations, insisting more on the separatrix case, which plays a prominent role in Sections 10.1 and 10.2.

An immediate identification leads to the following equations:

$$\begin{cases} F_0\left(x_1^0, x_2^0, \ldots, x_n^0\right) = C_0, \\ \sum n_i^0 \dfrac{\partial S_1}{\partial y_i} = 0, \\ \sum n_i^0 \dfrac{\partial S_2}{\partial y_i} + \dfrac{1}{2}\sum \dfrac{\partial^2 F_0}{\partial x_i \partial x_k}\dfrac{\partial S_1}{\partial y_i}\dfrac{\partial S_1}{\partial y_k} + F_1(x^0, y) = C_2, \\ \sum n_i^0 \dfrac{\partial S_3}{\partial y_i} + \dfrac{1}{2}\sum \dfrac{\partial^2 F_0}{\partial x_i \partial x_k}\dfrac{\partial S_1}{\partial y_i}\dfrac{\partial S_2}{\partial y_k} + \Phi_3 = 0, \\ \sum n_i^0 \dfrac{\partial S_4}{\partial y_i} + \dfrac{1}{2}\sum \dfrac{\partial^2 F_0}{\partial x_i \partial x_k}\dfrac{\partial S_1}{\partial y_i}\dfrac{\partial S_3}{\partial y_k} + \Phi_4 = C_4, \\ \ldots \end{cases}$$

where, following Poincaré, I have denoted by Φ_3, Φ_4 known functions and adopted conventions for the summations which the reader will guess easily.

From the hypothesis that $n_1^0 = 0$ is the sole linear relation with integer coefficients between the frequencies, we deduce from the second equation that S_1 is necessarily of the following form:

$$S_1 = \alpha_1 y_1 + \alpha_2 y_2 + \cdots + \alpha_n y_n + f(y_1),$$

where the derivative f' of f is periodic.

---

[97]This will be pertinent in the case of the Restricted Problem.

[98]By a symplectic change of coordinates one reduces any resonance relation $\sum m_i n_i^0 = 0,\ m_i \in \mathbb{Z}$, to $n_1 = 0$.

[99]After changing coordinates in the case of libration.

Given a function $U(y)$, all of whose derivatives $\frac{\partial U}{\partial y_i}$ are periodic,

$$U = \sum \alpha_i y_i + \sum_{p=(p_1,\dots,p_n)\in\mathbb{Q}^n} a_p \sin(\sum p_i y_i) + \sum_{p=(p_1,\dots,p_n)\in\mathbb{Q}^n} b_p \cos(\sum p_i y_i),$$

as Poincaré we let

$$[U] = \sum \alpha_i y_i + \sum_{p_1\in\mathbb{Q}} a_{(p_1,0,\dots,0)} \sin(p_1 y_1) + \sum_{p_1\in\mathbb{Q}} b_{(p_1,0,\dots,0)} \cos(p_1 y_1),$$

the sum of the secular and resonant terms in U. It follows from this definition that, for any such U,

$$\sum n_i^0 \frac{\partial [U]}{\partial y_i} = \sum n_i^0 \alpha_i$$

is a constant. Hence one deduces from the third equation, that (I follow Poincaré's notation for the constants):

$$\frac{1}{2}\left[\sum \frac{\partial^2 F_0}{\partial x_i \partial x_k}\frac{\partial S_1}{\partial y_i}\frac{\partial S_1}{\partial y_k}\right] = C_2' - [F_1](x^0, y),$$

which, given the expression of S_1, is an equation of the form

$$Af'^2 + 2Bf' + D = C_2' - [F_1](x^0, y),$$

the coefficients A, B, D depending on the α_i which we can choose as we please. Choosing all the α_i equal to 0, this implies $C_2' = C_2$ and the formula defining f' becomes:

$$Af'^2 = C_2 - [F_1](x^0, y).$$

The three cases are respectively

circulation: $C_2 > \max[F_1]$, libration: $C_2 < \max[F_1]$, separatrix: $C_2 = \max[F_1]$.

In the separatrix case, the formula $f' = \sqrt{A^{-1}(C_2 - [F_1])}$ defines two 4π-periodic functions of y_1.

This is only the beginning of the story which leads to proving the existence of the S_p, whose partial derivatives are 4π-periodic in y_1 and 2π-periodic in $y_2, \dots, y_n$. Indeed, some work is needed to show that the constants $C_4, \dots$ can be chosen in such a way that thc $\frac{\partial [S_p]}{\partial y_1}$ remain finite; this is not obvious because in the equation $\frac{\partial^2 F_0}{\partial x_1^2}(x^0)\frac{\partial S_1}{\partial y_1}\frac{[S_p]}{\partial y_1} = \cdots$, the term $\frac{\partial S_1}{\partial y_1}$ vanishes twice per period. Here is the way, at the end of Section 209, Poincaré describes the formal penduli defined by the solutions of separatrix type that he has just found for the Hamilton–Jacobi equation:

> Nous avons donc pu déterminer des fonctions satisfaisant aux conditions que nous nous étions imposées et nous avons réalisé une véritable *généralisation des solutions périodiques.* Seulement, tandis que les séries qui définissent les solutions périodiques sont convergentes, il n'en est plus de même de celles dont nous venons de démontrer l'existence, de

> sorte que cette généralisation n'a de valeur qu'au point de vue du calcul formel.[100]

This divergence is crucial for the dynamics; we shall come back to this point in Section 10.1.

Being mainly interested in the sequel in the Restricted Problem, I shall not discuss the degenerate case which occurs for the Non-Restricted Three-Body Problem. But, without surprise, Poincaré studies it in depth in Chapter XXI. Arguing that the Lindstedt series still exist in this case, he says:

> On peut donc prévoir que la méthode de M. Bohlin est encore applicable aux cas où F_0 ne dépend pas de toutes les variables de la première série et, en particulier, au problème des trois Corps. Mais l'application soulève quelques questions délicates et je suis obligé d'insister.[101]

8. Integral invariants and Poisson stability

We have already seen in Section 6.1 that, when studying the exponents of periodic solutions, Poincaré makes use of the symplectic structure of the equations. But it is in Chapter XXII, which opens the third volume of the New Methods that he displays the full consequences of this property.

8.1. Integral invariants as integrals of the variational equations

Having proved the non-existence of first integrals besides the classical ones which come from the symmetries of the problem, Poincaré sees the integral invariants as an ersatz: it consists in replacing the equations of motion by the *variational equations* which, indeed, admit first integrals. Explained in Section 242, this filiation is largely forgotten in modern expositions:

> Reprenons le système
>
> $$\frac{dx_1}{X_1} = \frac{dx_2}{X_2} = \cdots = \frac{dx_n}{X_n} = dt. \qquad (1)$$
>
> Nous pouvons former les équations aux variations correspondantes telles qu'elles ont été définies au début du Chapitre IV. Pour former ces équations, on change dans les équations (1) x_i en $x_i + \xi_i$ et l'on néglige les carrés des ξ_i ; on trouve ainsi le système d'équations linéaires
>
> $$\frac{d\xi_k}{dt} = \frac{dX_k}{dx_1}\xi_1 + \frac{dX_k}{dx_2}\xi_2 + \cdots + \frac{dX_k}{dx_n}\xi_n \,. \qquad (2)$$

[100] We have thus been able to determine functions which satisfy the conditions we had imposed and we have accomplished a true *generalization of the periodic solutions*. But, while the series defining the periodic solutions are convergent, this is no more true of those which we have just proved to exist, so that this generalization has value only from the point of view of formal computation.

[101] Hence one can predict that M. Bohlin's method is still applicable to the case where F_0 does not depend on all the variables of the first series, and, in particular, to the Three-Body Problem. But the applications raises some delicate questions and I am forced to insist.

> Il y a, entre les intégrales des équations (2) et les invariants intégraux des équations (1), un lien intime qu'il est aisé d'apercevoir.
> Soit $F(\xi_1, \xi_2, \cdots, \xi_n) =$ const., une intégrale quelconque des équations (2). Ce sera une fonction homogène par rapport aux ξ, et dépendant d'ailleurs des x d'une manière quelconque. Je pourrai toujours supposer que cette fonction F est homogène de degré 1 par rapport aux ξ ; car s'il n'en était pas ainsi, je n'aurais qu'à élever F à une puissance convenable pour trouver une fonction homogène de degré 1. Considérons maintenant l'expression
> $$\int F(dx_1, \cdots, dx_n),$$
> je dis que c'est un invariant intégral du système (1).[102]

and more specifically, in Section 255:

> Dans le cas des équations de la Dynamique, il est aisé de former un grand nombre d'invariants intégraux. Nous avons en effet appris, aux n°s 56 et suivants, à former un certain nombre d'intégrales de l'équation aux variations et nous avons appris dans le Chapitre précédent comment on peut en déduire des invariants intégraux.
>
> Une première intégrale (équation 3, t. I, p. 167) est la suivante
> $$\eta_1'\xi_1 - \xi_1'\eta_1 + \eta_1'\xi_1 - \xi_1'\eta_1 + \cdots = \text{const.}$$
> L'invariant intégral qu'on en déduit est le suivant
> $$J_1 = \int (dx_1 dy_1 + dx_2 dy_2 + \cdots dx_n dy_n).$$
> Il est du deuxième ordre et fort important pour ce qui va suivre.[103]

Then, writing determinants which amount to computing the exterior powers of the symplectic form $dx_1 \wedge dy_1 + dx_2 \wedge dy_2 + \cdots + dx_n \wedge dy_n$, he concludes:

---

[102] Let us take back the system (1) $\frac{dx_1}{X_1} = \frac{dx_2}{X_2} = \cdots = \frac{dx_n}{X_n} = dt$. We can form the corresponding variational equations as they were defined at the beginning of Chapter IV. In order to form these equations one changes x_i into $x_i + \xi_i$ in equations (1) and one neglects the squares of the ξ_i ; one finds in this way the system of linear equations (2) $\frac{d\xi_k}{dt} = \frac{dX_k}{dx_1}\xi_1 + \frac{dX_k}{dx_2}\xi_2 + \cdots + \frac{dX_k}{dx_n}\xi_n$. There is, between the integrals of equations (2) and the integral invariants of equations (1), an intimate relation which is easy to see. Let $F(\xi_1, \xi_2, \ldots, \xi_n) =$ const., be any integral of equations (2). It is a function, homogeneous of degree 1 in the ξ and depending in an arbitrary way on the x; indeed, if this was not the case, it would be enough to raise F to a conveniently chosen power in order to obtain a function homogeneous of degree 1. Now, let us consider the expression $\int F(dx_1, \ldots, dx_n)$, I say it is an integral invariant of system (1).

[103] In the case of the equations of Dynamics, it is easy to form a large number of integral invariants. Actually, we have learned in n°s 56 and the following, to form a number of integrals of the variational equation and we have learned in the previous chapter how to deduce from them integral invariants. A first integral (equation 3, t. I, p. 167) is the following $\eta_1'\xi_1 - \xi_1'\eta_1 + \eta_1'\xi_1 - \xi_1'\eta_1 + \cdots =$ const.. The resulting integral invariant is the following $J_1 = \int(dx_1dy_1 + dx_2dy_2 + \cdots + dx_ndy_n)$. It is of second order and very important for what follows.

> Cependant, parmi tous ce invariants, il y en a un auquel il convient d'attacher une grande importance, c'est le dernier d'entre eux[104]
>
> $$J_n = \int dx_1 dy_1 dx_2 dy_2 \cdots dx_n dy_n.$$

This is of course the *Liouville theorem*, that is the preservation by a Hamiltonian flow of the volume in phase space. Moreover, the interior product of the symplectic form by a particular solution of the variational equation furnishes other invariants: namely, if $\xi = \varphi(t), \eta = \psi(t)$ is a solution of the variational equation, the constancy of $\xi\psi - \eta\varphi$ for any other solution (ξ, η), implies that $\int(\psi dx - \varphi dy)$ is a (time dependent) integral invariant. In addition to the ones coming from the derivation of actual integrals of the equations of motion, Poincaré did not fail to notice that the homogeneity of the potential gives such a particular solution: if $q(t) = (\vec{r}_1(t), \vec{r}_2(t), \vec{r}_3(t))$ is a solution of the Three-Body Problem, so is $\lambda^2 q(\frac{t}{\lambda^3})$ for any $\lambda > 0$; taking the derivative with respect to λ at $\lambda = 1$ gives the required solution. This invariance under rescaling is at the root of the analysis of the behaviour of solutions of the Three-Body Problem near a triple collision, first by Sundman and then more geometrically by R. McGehee thanks to the introduction of a compactification called the *collision manifold* (see [McG, C7]). It is only in the case of the *strong force potential*, proportional to the inverse square of the mutual distance that this scaling property becomes a true symplectic symmetry of the equations, implying the existence of the *Jacobi integral*[105] (see [AC]).

But there is more: in Section 261, entitled "The integral invariants and the characteristic exponents", Poincaré deepens the analogy between integrals of the equations of motion and integral invariants, by asking:

> On peut se demander s'il existe d'autres invariants intégraux algébriques que ceux que nous venons de former.
>
> On pourrait appliquer, soit la méthode de Bruns, soit celle dont j'ai fait usage aux chapitres IV et V ; en effet, les invariants intégraux correspondent, comme nous l'avons vu, aux intégrales des équations aux variations et l'on pourrait appliquer à ces équations les mêmes procédés qu'aux équations du mouvement elles-mêmes.[106]

Even if the analysis is not complete, because of the difficult control of the genericity of the periodic solutions, it has the interest of indicating probable bounds on the number of integral invariants. Poincaré concludes:

---

[104] However, among all those invariants, there is one to which one should attach a great importance, it is the last of them $J_n = \int dx_1 dy_1 dx_2 dy_2 \cdots dx_n dy_n$.

[105] Poincaré had read Jacobi's *Vorlesungen über Dynamik.*

[106] One could ask whether there exists algebraic invariants different from those which we have just formed.

One could apply, either Bruns's method, or the one which I used in Chapters IV and V; indeed, as we saw, the integral invariants correspond to the integrals of the variational equations and one could apply to these equations the very method we applied to the equations of motion themselves.

> Il est probable que ces invariants nouveaux, dont la discussion précédente n'exclut pas la possibilité, n'existent pas ; mais pour le démontrer, il faudrait recourir à d'autres procédés, par exemple à des procédés analogues à la méthode de Bruns.[107]

To my knowledge, this has not been pursued explicitly, but see [Mal, Cas].

Remarks

1) In the introduction of his first book *Leçons sur les invariants intégraux* [Ca], directly inspired by the works of Poincaré, Elie Cartan writes:

> En définitive, *la quantité sous le signe d'intégration dans un invariant intégral de H. Poincaré n'est autre chose qu'une forme différentielle invariante tronquée.* Le caractère invariant de l'intégrale *complétée* est conservé si elle est étendue à un ensemble quelconque d'états, *simultanés ou non.*[108]

The paradigmatic example of what Cartan calls a "completion" is the transformation of the *Poincaré integral invariant* $\int pdq$ into the *Poincaré–Cartan integral invariant* or, in Cartan's terminology[109], *impulsion-energy tensor* $\int (\sum_i p_i dq_i - H(p,q,t)dt)$; both are *relative invariants*, which means that they have to be integrated on closed loops: the first integral does not change by transport of the loop along the flow lines, either between two fixed instants or at fixed energy; the second one is invariant by any transport of the loop along graphs of flow lines in the (q,p,t) space. It appears already in Section 336 which opens Chapter XXIX; studying the variations of the integral[110] $J = \int_{t_0}^{t_1} \left(-F + \sum y_i \frac{dx_i}{dt}\right)$, Poincaré, writes:

> Nous avons supposé jusqu'à présent que les deux limites t_0 et t_1 sont données ; qu'arrive-t-il si les limites sont regardées comme variables ?[111]

Supposing the Hamiltonian is autonomous and noticing that the partial derivative of the action with respect to a boundary of the time interval is, up to sign, the energy, he concludes:

> Si cette constante [l'énergie] est nulle, l'action J est encore minimum si l'on regarde les valeurs initiales et finales des variables x_i comme données

[107] It is likely that these new invariants, whose possibility is not excluded by the preceding discussion, do not exist; but in order to prove it, we need to resort to other methods, e.g., processes similar to the method of Bruns.

[108] Ultimately, the quantity under the sign of integration in an integral invariant of H. Poincaré is nothing but a truncated invariant differential form. The invariant character of the completed integral is preserved if it is extended to any set of states, simultaneous or not.

[109] Physicists rather speak of the impulsion-energy quadri-(co)vector. Anyway, this is just the action integral.

[110] that is, up to sign, the integral of $\sum p_i dq_i - Hdt$.

[111] Up to now, we have supposed that both limits t_0 and t_1 are given; what happens if these limits are considered as variable?

> et *quand même on ne regarderait pas comme données les valeurs initiales et finales du temps, t_0 et t_1*.[112]

Some pages later, in Section 341, he will correct this statement in his quite charming characteristic style:

> Jusqu'ici, quand j'ai dit, *telle intégrale est minimum*, je me suis servi d'une façon de parler abrégée, mais incorrecte, qui ne pouvait d'ailleurs tromper personne ; je voulais dire, *la variation première de cette intégrale est nulle* ; cette condition est nécessaire pour qu'il y ait minimum, mais elle n'est pas suffisante.[113]

Finally, recall that the property of the impulsion-energy tensor of being a relative integral invariant characterizes the equations of Hamilton (see the introduction of [Ca] or [A2]): it means that the graphs of the flow lines are tangent to the one-dimensional kernel of the exterior derivative of $pdq - Hdt$.

2) Continuing with the introduction of [Ca], one finds the strong link between the integral invariants in the sense of Poincaré and Cartan and the ones in the sense of Lie, directly relevant to the theory evoked in 6.5:

> La notion d'invariant intégral peut être envisagée d'un point de vue un peu différent du point de vue habituel qui est celui de Poincaré, et qui est en somme celui où on s'est plaçé dans ces Leçons. Au lieu de considérer une intégrale multiple attachée à un système d'équations différentielles vis-à-vis duquel elle jouit d'une propriété d'invariance, on peut la considérer comme attachée à un groupe de transformations par rapport auquel elle est invariante. Les deux points de vue sont du reste connexes. Le dernier est celui auquel s'est placé S. Lie et qui lui a paru pendant quelque temps le seul vrai. Là encore la notion d'invariant intégral joue un rôle important puisque, comme l'auteur l'a montré [*Sur la structure des groupes infinis de transformations, Annales de l'E.N.S.* 1904 1905], tout groupe de transformations peut, au besoin par l'adjonction de variables auxiliaires, être défini comme l'ensemble des transformations qui admettent un certain nombre d'invariants intégraux linéaires.[114]

---

[112] If this constant [the energy] is equal to zero, the action J is still minimum if one considers the initial and final values of the variables x_i as given and *even if one would not consider as given the initial and final values of the time, t_0 and t_1*.

[113] So far, when I said, *this integral is minimum*, I used a shorthand, but incorrect, way of talking, but one which could not mislead anyone; I meant, *the first variation of this integral is zero*; this condition is necessary for being a minimum, but it is not sufficient.

[114] The notion of integral invariant can be envisaged from a somewhat different point of view of the usual one which is the one of Poincaré, and which is basically the one which we have taken in these lessons. Instead of considering a multiple integral attached to a system of differential equations with respect to which it enjoys an invariance property, it can be considered as attached to a group of transformations with respect to which it is invariant. Indeed, the two views are related. The latter is the one of S. Lie who, for a while, considered it as the only true one. Here too, the notion of integral invariant plays an important role since, as the author showed [*On the structure of infinite groups of transformations, Annals of the ENS* 1904 1905], any group

8.2. Poisson stability

> Le mot stabilité a été entendu sous les sens les plus différents, et la différence de ces divers sens deviendra manifeste si l'on se rappelle l'histoire de la Science. Lagrange a démontré qu'en négligeant les carrés des masses, les grands axes des orbites deviennent invariables. Il voulait dire par là qu'avec ce degré d'approximation les grands axes peuvent se développer en séries dont les termes sont de la forme $A\sin(\alpha t+\beta)$, A, α et β étant des constantes.[115]

The title of Chapter XXVI is not innocent. Having wrongly stated in the original version of the memoir in 1889 a very strong stability theorem in the Planar Circular Restricted Three-Body Problem, Poincaré needs a stability result which, if much weaker, will turn to be of considerable importance.[116] Indeed, based on the *recurrence theorem*, it is the forerunner of *ergodic theory*. The *Poisson stability* alludes to the absence of purely secular terms (i.e., of terms which grow without limits when times goes on) in the planetary semi-major axes at the second order of the classical theory of perturbations (i.e., when neglecting the cubes of the planetary masses) which, at this order of approximation, implies a recurrent behaviour of these semi-major axes (Spiru Haret will show that this property is not shared any more by the approximations of higher orders.) Using an argument of conservation of the volume in a container of finite volume, Poincaré shows that, in the case he is considering, the conservation of the integral invariant implies that a "generic" solution of the Restricted Problem will come back infinitely often in an arbitrarily small neighborhood of a given point of the phase space. In the following quotation (Chapter XXVI Section 296) Poincaré legitimates his new use of the word "stability":

> En résumé, les molécules qui ne traversent U_0 qu'un nombre fini de fois sont exceptionnelles au même titre que les nombres commensurables qui ne sont qu'une exception dans la série des nombres, pendant que les nombres incommensurables sont la règle. Si donc Poisson a cru pouvoir répondre affirmativement à la question de la stabilité telle qu'il l'avait posée, bien qu'il eût exclu les cas où le rapport des moyens mouvements est commensurable, nous aurons de même le droit de regarder comme démontrée la stabilité telle que nous la définissons, bien que nous soyons

of transformations may, after adding auxiliary variables if necessary, be defined as the set of transformations that admit some linear integral invariants.

[115]The word stability has been taken in very different meanings, and the difference between these meanings will become obvious if one remembers history of Science. Lagrange has proved that if one neglects the squares of the masses, the major axes of the orbites become invariable. He wanted to say that with this degree of approximation the major axes can be expanded in series whose terms are of the form $A\sin(\alpha t+\beta)$, where A, α and β are constants.

[116]In the first version of the Memoir, Poincaré deduced this weaker stability from the actual stability that he thought he had proved.

> forcés d'exclure les molécules exceptionnelles dont nous venons de parler.[117]

Titled *Probabilities*, this Section 296 has a striking visionary character: disregarding the fears of Joseph Bertrand about the "paradoxes" of continuous probabilities, Poincaré understands clearly that any choice of a regular density will lead to the same notion of negligible sets; in other words, he understands that the notion of *zero measure set* does not depend on such a choice:

> Mais il faut d'abord que j'explique le sens que j'attache au mot *probabilité*. Soit $\varphi(x, y, z)$ une fonction quelconque positive des trois coordonnées x, y, z ; je conviendrai de dire que la probabilité pour qu'à l'instant $t = 0$ une molécule se trouve à l'intérieur d'un certain volume est proportionnelle à l'intégrale
>
> $$J = \int \varphi(x, y, z) dxdydz$$
>
> étendue à ce volume. ... Nous pouvons choisir arbitrairement la fonction φ et la probabilité se trouve ainsi complètement définie. $\cdots$ Nous retombons donc sur les mêmes résultats *qui sont ainsi indépendants du choix de la fonction* φ.[118]

A beautiful analysis of this part of Poincaré's text can be found in the thesis of Anne Robadey [Ro]. See also the pleasant paper [Gh] to get acquainted with possible misinterpretations of the recurrence theorem.

8.3. From the recurrence theorem to ergodic theory

The recurrence theorem is a weak stability result but it tells us nothing on the way the solution will come back indefinitely often close to its initial point. It could be in an integrable way, staying on an invariant torus and hence exploring a very small part of the energy hypersurface or else in an *ergodic* way, filling a dense subset in the energy surface. The ergodic theorem makes this difference quantitatively precise.

---

[117]Summarizing, the molecules which pass through U_0 only a finite number of times are exceptional in the same way as the rational numbers which are but an exception in the series of numbers, while the irrational numbers are the rule. Hence if Poisson considered that he could reply affirmatively to the question of the stability as he had set it, although he had excluded the case when the mean motions are commensurable, we shall have in the same way the right of considering the stability as proved, although we are forced to exclude the exceptional molecules that we have just mentioned.

[118]But I must first explain the meaning I give to the word *probability*. Let $\varphi(x, y, z)$ any positive function of the three coordinates x, y, z; I shall agree to say that the probability for a molecule to be at time 0 inside a certain volume is proportional to the integral

$$J = \int \varphi(x, y, z) dxdydz$$

extended to this volume. ... We may choose arbitrarily the function φ and the probability is then completely defined. ... So doing, we get again the same results *which hence are independent of the choice of the function* φ.

The notion of ergodicity originates from the works of Boltzmann in statistical mechanics, namely his famous *Theorem H* (1872) which asserts the monotonicity of the entropy in a system formed by a great number of interacting particles, as for example the atoms in a gas. As noticed by Poincaré and Zermelo, this appears to be in contradiction with Poincaré's recurrence theorem. Here is what Poincaré writes in a short paper titled *Le mécanisme et l'expérience*, which appeared in 1893 in the *Revue de métaphysique et de morale*:

> Un théorème facile à établir nous apprend qu'un monde limité soumis aux seules lois de la mécanique, repassera toujours par un état très voisin de son état initial. Au contraire, d'après les lois expérimentales admises (si on leur attribue une valeur absolue et qu'on veuille en pousser les conséquences jusqu'au bout), l'univers tend vers un certain état final dont il ne pourra plus sortir. Dans cet état final, qui sera une sorte de mort, tous les corps seront en repos et à la même température...
>
> ... Cet état ne sera donc pas une mort définitive de l'univers, mais une sorte de sommeil, d'où il se réveillera après des millions de millions de siècles. A ce compte, pour voir la chaleur passer d'un corps froid à un corps chaud, il ne serait plus nécessaire d'avoir la vue fine, la présence d'esprit, l'intelligence et l'adresse du démon de Maxwell, il suffirait d'un peu de patience.
>
> On voudrait pouvoir s'arrêter à cette étape et espérer qu'un jour le télescope nous montrera un monde en train de se réveiller et où les lois de la thermodynamique seront renversées.
>
> Malheureusement d'autres contradictions surgissent...[119]

Boltzmann replies that the time of recurrence being a priori greater than, say, the age of the solar system, this close return is never witnessed in practice for a system as complex as a gas.

For the apparent contradiction, raised by Poincaré and Zermelo, between Boltzmann's theorem or Landau damping and the recurrence theorem, see [V] Section 10. The main arguments are the infinite number of degrees of freedom and the infinite limit of the recurrence time when the number of particles goes to infinity, the importance of the preparation of the initial condition, ...

---

[119] A theorem, easy to establish, tells us that a bounded world, only subject to the laws of mechanics, will always revert to a state very close to the original one. Instead, according to the admitted experimental laws (if one assigns to them an absolute value and if we want to push through the consequences), the universe tends toward a final state from which it will not be able to escape. In this final state, which is a kind of death, all bodies are at rest and at the same temperature...

... This condition will not be final death of the universe, but a kind of sleep, from which it will wake up after millions of millions of ages. If so, in order to see the heat move from a cold body to a hot body, it would no longer be necessary to have the sharp view, the presence of mind, the intelligence and the address of Maxwell's demon, a little patience would suffice.

One would wish to be able to stop at this point and hope that, one day, the telescope shows us a world waking up and where the laws of thermodynamics will be overturned.

Unfortunately other contradictions arise....

The mathematical theory, named *ergodic theory*, comes to maturity in 1930, with the ergodic theorems of Von Neumann, Birkhoff and Koopman, which are strong dynamical forms of the *strong law of large numbers*. It goes beyond Poincaré's *recurrence theorem* in that it makes precise the relation between averages in the phase space and time averages in a dynamical system preserving a measure whose total mass is finite (which one can normalize to a probability measure, that is one with total mass equal to 1). More precisely, a system is said to be *ergodic* if the measure of any invariant subset is necessarily equal to 0 or 1, and a consequence of Birkhoff's ergodic theorem is that the mean recurrence time of almost every point in a measurable subset A is inversely proportional to the measure of A. A good introduction is the little book by Sinai [Sin].

9. Stroboscopy 1) Planar Circular Restricted Three-Body Problem

9.1. The simplest Three-Body Problem

> En ce qui concerne le problème des trois corps, je ne suis pas sorti du cas suivant : Je considère trois masses, la première très grande, la seconde petite mais finie, la troisième infiniment petite ; je suppose que les deux premières décrivent un cercle autour de leur centre de gravité commun et que la troisième se meut dans le plan de ces cercles. Tel serait le cas d'une petite planète troublée par Jupiter, si l'on négligeait l'excentricité de Jupiter et l'inclinaison des orbites.[120]

This vivid description of the so-called *Planar Circular Restricted Three-Body Problem* in the planetary case introduces the main theme of the memoir *On the Three-Body Problem and the equations of Dynamics* which, as we already mentioned, wins in 1889 the prize of the King of Sweden, and makes its 35 years old author known in large circles. The small parameter μ is ratio of the mass of Jupiter to the mass of the Sun, that is approximately 1/1000. Similar equations rule the motions of the couple Earth-Moon perturbed by the Sun, the small parameter depending now on the ratio of distances (in fact semi-major axes) $r/R = d(\text{Earth,Moon})/d(\text{Earth,Sun})$ (more accurately, on the ratio $\sqrt{m_S/m_E}(r/R)^{3/2}$ of the mean motion frequencies, that is approximately 30/365). G.W. Hill's works on this last problem strongly influenced Poincaré. In a rotating frame fixing the couple Sun-Jupiter in the planetary case, the couple Earth-Sun in the Lunar case, the equations have the same canonical form as above, F being now the *Jacobi constant*. With the *geodesic flow* on an almost spherical surface, which Poincaré studies at the end of his life, these

[120]Concerning the Three-Body Problem, I did not venture away from the following case: I consider three masses, the first one very big, the second one small but finite, the third one infinitesimal; I suppose that the two first describe a circle around their common center of mass and that the third moves in the plane of these circles. Such would be the case of a small planet disturbed (*I hope this word may convey the charm of the eighteen century flavoured "troublé"*) by Jupiter if one was neglecting Jupiter's eccentricity and the inclination of the orbits.

are natural examples of non-integrable Hamiltonian systems with $N = 2$ degrees of freedom.

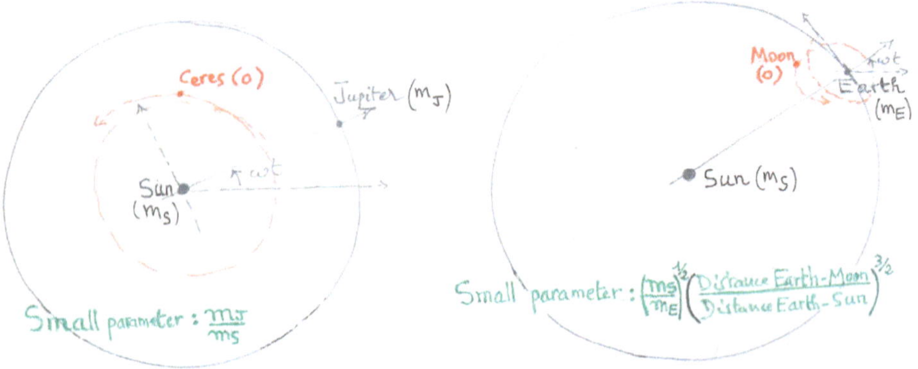

PLANETARY CASE LUNAR CASE

FIGURE 10. The Restricted Three-Body Problem.

In *canonical heliocentric*[121] *coordinates*, the Hamiltonian (= Jacobi constant) takes the form

$$F(L, G, l, g) = -\frac{K}{L^2} + G + \mu F_1(L, G, l, g - t),$$

where K is a constant and $g - t$ is the argument of the perihelion of the Keplerian ellipse in the frame which rotates with angular velocity 1 (as assumed by Poincaré).

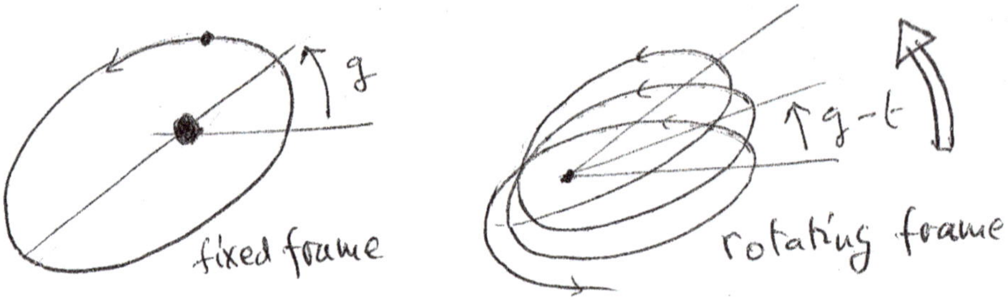

FIGURE 11

One nice feature is that, if still degenerate (the Hessian of the integrable part $F_0(L, G) = \frac{K}{L^2} + G$ vanishes), F is easily transformed into a non-degenerate Hamiltonian by Poincaré's trick of replacing it by its exponential $expF$, which has the same integrability and stability properties. This replacement affects only the law of time; in particular,

the return map (*see below Section* 9.3) *is the same and hence non-degenerate.*

[121] in the planetary case, canonical geocentric in the lunar case.

Finally, as we already said, Delaunay coordinates are not defined in the neighborhood of circular motions and Poincaré replaces them by

$$x_1 = L - G,\ x_2 = L + G,\ y_1 = \frac{1}{2}(l - g + t),\ y_2 = \frac{1}{2}(l + g - t),$$

which are in reality 2–1 (a fact noticed by Poincaré in the Memoir but forgotten by him in *The New Methods*). A last step, considering these as symplectic polar coordinates, leads to the *Poincaré coordinates* (see Section 2.2) $\sqrt{2x_1}e^{iy_1}, \sqrt{2x_2}e^{iy_2}$, which are regular at the circular direct or retrograde Kepler orbits. There are still problems near the collision orbits but Poincaré is only interested in the neighborhood of the circular ones.

9.2. Hill stability

Stability is the main question addressed by Poincaré in his Memoir (but not the main question asked for the prize, which was to find convergent series expansions of the coordinates of the bodies in case no collision occurs [BG]).
For the 2 degrees of freedom Restricted Problem, this is a question a priori simpler than for the general Three-Body Problem (see the paragraphs on K.A.M. and diffusion). Indeed, Hill had already proved that some kind of stability occurs when the Jacobi constant is large enough, as shown on Figure 12 where an energy hypersurface and its projection on the configuration space (delimiting the so-called *Hill's regions*) are drawn, for such values of the Jacobi constant: escape of the infinitesimal mass from the Hill region is impossible but collisions with the closest primary are not excluded.

9.3. Return map

In a rotating frame, the planar circular Three-Body Problem becomes an autonomous Hamiltonian system with two degrees of freedom. Fixing the energy (= *Jacobi constant*) to a high enough value and keeping only the part of the three-dimensional energy hypersurface above the bounded Hill region around the Sun (planetary case) or the Earth (Lunar case), one gets after regularization of the collisions,[122] a manifold diffeomorphic to $SO(3)$, that is to the real projective space $\mathbb{R}P^3$. Poincaré writes down explicit coordinates in the 2-fold covering, diffeomorphic to S^3 of such a hypersurface deprived of one point, which makes it diffeomorphic to $\mathbb{R}^3$.

Being interested only in motions of the 0-mass body close to circular (direct or retrograde), Poincaré does not care about the fact that his coordinates are not smooth at collisions.[123]

[122]The regularization, due to Levi-Civita [LC] (using a complex transformation anticipated by Goursat [Go] in the same year 1889 as the printing of Poincaré's Memoir), amounts to compactifying the energy submanifold by the addition of a circle corresponding to every possible direction of the velocities of the colliding bodies (see [Co]).
[123]because l must be defined from the eccentric anomaly u by Kepler's formula $l = u - e\sin u$.

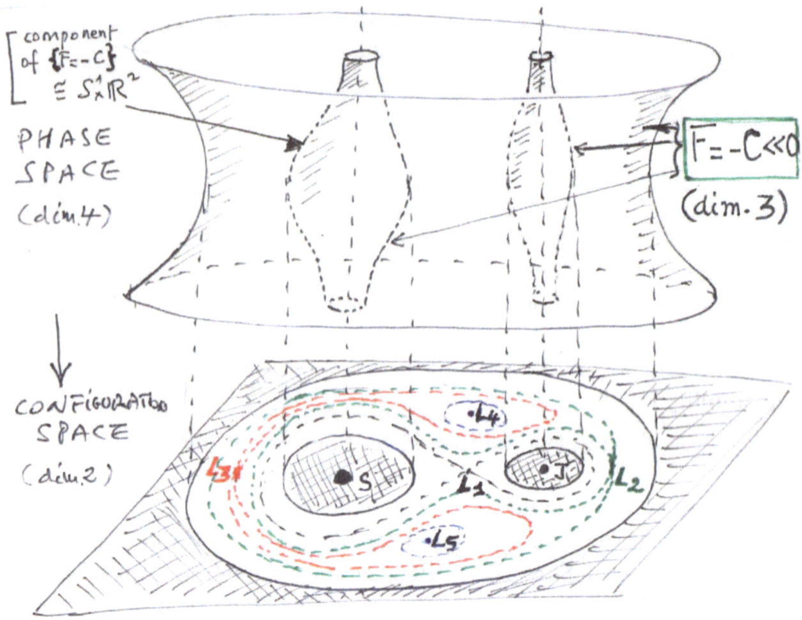

FIGURE 12. Hill stability.

In Section 305 of Chapter XXVII, Poincaré embeds in this $\mathbb{R}^3$ a half-plane with the property that all integral curves, except the retrograde *Hill orbit*,[124] cut it transversely an infinite number of times. Such a half-plane thus becomes a *stroboscope*; describing the succession $M_0, M_1, \ldots$ of points of intersection of a single trajectory, Poincaré writes:

> Le point M_1 sera dit le *conséquent* de M_0. Ce qui justifie cette dénomination, c'est que, si l'on considère le faisceau des courbes qui satisfont aux équations différentielles (1) ; si, par le point M_0, on fait passer une courbe et qu'on la prolonge jusqu'à ce qu'elle rencontre de nouveau le demi-plan $(y = 0, x > 0)$, cette nouvelle rencontre aura lieu en M_1.[125]

Figure 14 shows the surfaces of section and the return maps, both in the unperturbed $(F_0(L,G) = -\frac{K}{L^2}+G)$ and perturbed $(F_0(L,G) = -\frac{K}{L^2}+G+\mu F_1(L,G,l,g-t))$ cases.

---

[124]The direct and retrograde *Hill orbits* which constitute its boundary, are periodic solutions of the first sort which respectively continue the circular, direct and retrograde, Kepler orbit which still exist in the rotating frame.

[125]The point M_1 will be called the *consequent* of M_0. The justification of this denomination is that, if one considers the set of curves satisfying the differential equations (1); if, through the point M_0, passes a curve and if we extend this curve until it meets again the half-plane $(y = 0, x > 0)$, this new encounter will take place in M_1.

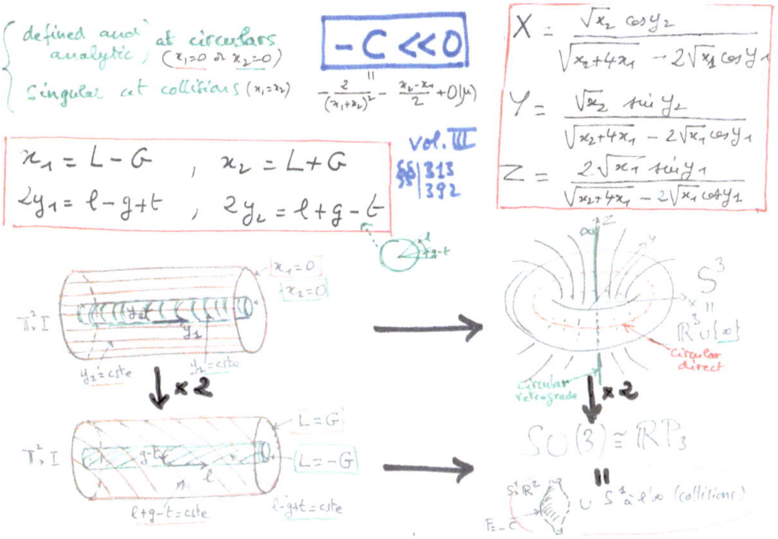

FIGURE 13. Coordinates in the compactified energy surface.

Hence the whole dynamics of the Restricted Problem with high Jacobi constant in a bounded Hill region is reduced to the study of the orbits under iteration of a mapping (the *Poincaré return map*) on a *surface of section* (here a half-plane but more accurately an annulus as described below; this return map may be loosely interpreted as describing the successive positions of the perihelion (or aphelion, which makes easier drawing, see Figure 14) of the osculating Keplerian ellipse of the zero mass body.
In other words, stroboscopy in an energy manifold has turned the continuous dynamical system defined by the equations of motion into a *discrete dynamical system*, i.e., the continuous time into a discrete time. Doing so, one looses the time parametrization of the solutions but not their asymptotic behaviour: periodic solutions become periodic orbits of the return map, quasi-periodic solutions dense in invariant tori become quasi-periodic orbits dense in invariant curves, ... Finally, *the question of stability can be wholly understood on the return map.*

Remark. Topologically, it is nicer to put the direct and retrograde Hill orbits on equal footing, and work in the *Birkhoff annulus of section* depicted on Figure 15. It is obtained from Poincaré half-plane by blowing up the fixed point corresponding to the direct Hill orbit and adding the missing point on the retrograde Hill orbit. It is embedded in S^3, which is a two-fold cover of the regularized energy manifold; it cuts transversely every solution, except the Hill orbits which are its boundary. Actually, the flow lines defined in a regularized energy surface by the (non-rotating) Kepler Hamiltonian $F_{Kep} = -\frac{K}{L^2}$ coincide with the fibers of a *Hopf fibration*: $S^3 \to S^2$, where the basis is the space of oriented ellipses (possibly degenerate)

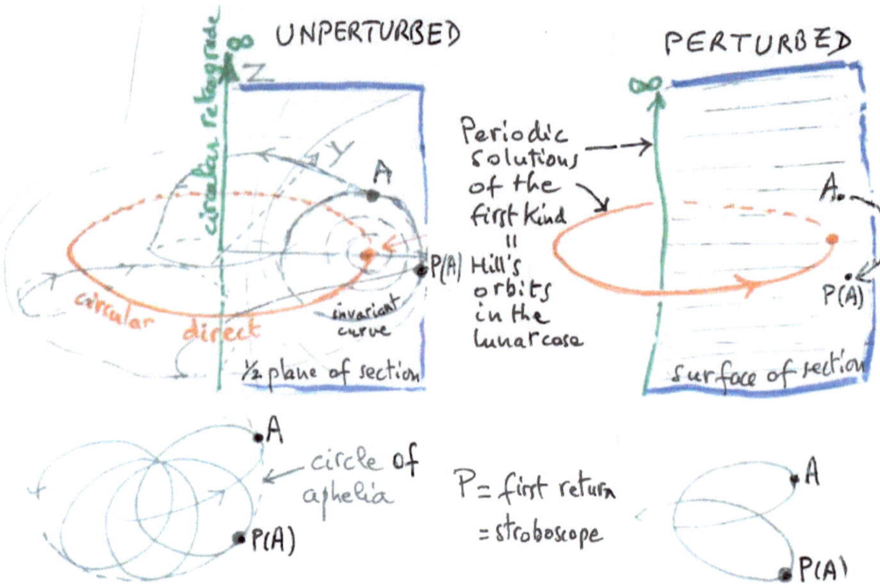

FIGURE 14. Poincaré return map.

with a fixed value of the semi-major axis. Blowing up two fibers (here the Hill orbits) is the simplest ersatz of a global section for this non-trivial fiber bundle. For more details, see [Co, C5]

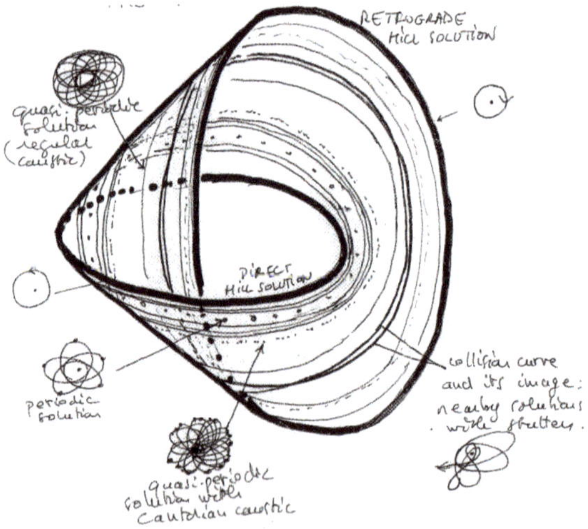

FIGURE 15. Birkhoff's annulus of section. Reproduced from [C5] with permission of © Springer 2008.

10. Resonances 2) Homoclinic and heteroclinic tangles

10.1. Divergence of the Bohlin series and exponentially small splitting

Let us stick to the Restricted Problem. The *periodic solutions of the second sort* result from the breaking of invariant tori of the unperturbed system F_0 filled with periodic solutions. In the surface of section, they correspond to periodic orbits bifurcating from invariant curves of the unperturbed return map filled with periodic orbits; they are in general born in pairs elliptic-hyperbolic when the parameter μ takes a non-zero value. At first order, the formal Bohlin series describe separatrices of the hyperbolic ones forming penduli and it was precisely the error of Poincaré to have considered these penduli as real (i.e., the Bohlin series to be convergent) and hence separating in the three-dimensional energy hypersurface. In the corrected version of the Memoir and in the second and third volume of the *New Methods*, Poincaré studies the actual phenomenon: if the stable and unstable invariant manifolds coincide formally (i.e., at any order of the theory of perturbations), as shown by the existence of formal Bohlin series in the separatrix case, they do not actually coincide in general.

In Chapter XXI (volume II), Poincaré shows by a direct computation the divergence of the Bohlin series in the case of a periodically forced pendulum. The equation is turned into a two degrees of freedom Hamiltonian by introducing the symplectic pair of energy-time (E, τ) coordinates:

$$\mathcal{F}(x, E, y, \tau) = E + F(x, y, \tau), \quad F(x, y, \tau) = x^2 - 2\mu \sin^2 \frac{y}{2} - \mu\epsilon\varphi(y) \cos \tau.$$

The same trick of decoupling the pendulum term and the perturbation term by the introduction of a second parameter will be used by Arnold in his famous example of diffusion [A4] to overcome the so-called "big gap" problem. For $\epsilon = 0$, the 2π-periodic solution $x = E = 0, y = 0, \tau = t$ has asymptotic solutions of the form

$$E = 0, \; x = \pm\sqrt{2\mu} \sin \frac{y}{2}, \; \tan \frac{y}{4} = Ce^{\pm t\sqrt{2\mu}}.$$

Seeking solutions $S(y, \tau) = S_0 + \epsilon S_1 + \epsilon^2 S_2 + \cdots$ of the Hamilton–Jacobi equation

$$F(\frac{\partial S}{\partial y}, y, \tau) = C = \epsilon C_1 + \epsilon^2 C_2 + \cdots,$$

we find

$$S_0 = \mp 2\sqrt{2\mu} \cos \frac{y}{2},$$

while S_1 must satisfies the equation

$$\frac{\partial S_1}{\partial \tau} + \frac{\partial F_0}{\partial x}\left(\frac{\partial S_0}{\partial y}\right)\frac{\partial S_1}{\partial y} - \mu\varphi(y) \cos \tau = C_1,$$

that is

$$\frac{\partial S_1}{\partial \tau} + (2\sqrt{2\mu} \sin \frac{y}{2})\frac{\partial S_1}{\partial y} - \mu\varphi(y) \cos \tau = C_1,$$

which is easily integrated by variation of the constants. Writing S_1 as the real part of a complex function and choosing complex paths to compute the relevant integrals, Poincaré finds two solutions $S_1(y,\tau)$ and $S_1'(y,\tau)$ with the property that S_1 vanishes for $y=2\pi$ and S_1' vanishes for $y=0$. Moreover, he evaluates the difference $S_1'-S_1$ by a simple residue computation; in the case where $\varphi(y)=\sin y$, the computation is explicit; it shows that

$$S_1'-S_1=O\left(\frac{1}{\sqrt{\mu}}e^{-\frac{\pi}{4\sqrt{2\mu}}}\right),$$

while exponentially small with respect to μ, is different from 0, which shows the divergence of the Bohlin series obtained by expanding S in powers of $\sqrt{\mu}$. For an interpretation in terms of the so-called Poincaré–Melnikov integral whose non vanishing is equivalent to the non vanishing of $S_1'-S_1$, see for example Chapter 7 of [AKN].

At the very end of the second volume of *The New Methods*, making the link with his study of the asymptotic solutions in Chapter VII (see Section 6.1), Poincaré gives a more geometric interpretation of this exponentially small difference in terms of the stable and unstable manifolds of the family of unstable (hyperbolic) periodic solutions which are the continuation of the one for $\epsilon=0$. These invariant manifolds do not coincide but they share a common formal asymptotic expansion in powers of $\sqrt{\mu}$ and this expansion is precisely the Bohlin series. For a thorough study of this example in the light of Ecalle's theory of *resurgent functions*, see [Sau]; for a more general survey, see [Ra2].

10.2. "... cette figure, que je ne cherche même pas à tracer"

Poincaré's mistake in the first version of the Memoir (see [BG, Y]) was asserting that the Bohlin series did converge and hence that these asymptotic curves formed penduli and hence true barriers to the dynamics:

> Donc *les surfaces asymptotiques sont des surfaces fermées*. Mais au début de ce travail, nous avons montré que pour établir la stabilité, il suffit de démontrer l'existence de surfaces trajectoires fermées.[126]

The discovery by Poincaré of the geometrical complexity caused by the non-convergence of the Bohlin series shapes our present understanding of the qualitative behaviour of solutions of a differential equation. In the following description, I combine Chapters XXVII and XXXIII. Here are the main steps:

1) Showing that the preservation of the symplectic form implies the preservation by the return map of a measure with a smooth density with respect to Lebesgue and finite total mass.[127] This is done by using explicit formulæ in the coordinates described in Figure 13.

2) Proving that the *asymptotic curves* which are the intersection with the surface of section of the stable and unstable manifolds of the newborn periodic

[126] *Hence the asymptotic surfaces are closed surfaces.* But at the beginning of this work, we have shown that in order to establish the stability, it is enough to prove the existence of closed surface-trajectories.

[127] Of course, Poincaré does not use these words because *Measure theory* does not exist yet.

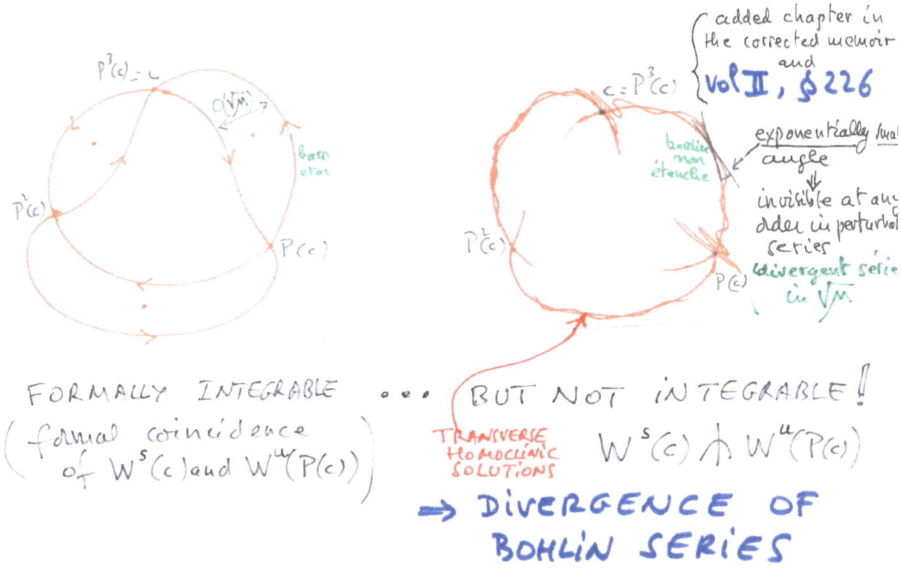

FIGURE 16. The error in the first version of the Memoir.

solutions must intersect (in Figure 17: A_0A_5 and B_0B_5 must intersect); the argument is a beautiful combination of invariance of the measure, topology of the plane and clever use of the fact that the asymptotic curves are exponentially close with respect to the parameter in a certain region, more precisely, that the segments A_0B_0 and A_5B_5 in Figure 17 are exponentially small.

The upshot is that the asymptotic curves must intersect and at the intersection points, they make exponentially small angles. It seems at first sight that piecing together segments of asymptotic curves, one defines closed curves which could form barriers implying stability of the dynamics as penduli would have done (see Figure 16) but it is not so. Such curves are not invariant and if iterated, they become an incredibly complicated tangle which is not a hermetic barrier but which nevertheless slows down the dynamics: one can get through but only very slowly, hence there is a priori no stability but the instability needs a very long time to be detectable.

Understanding the extreme complexity of the intersections of these two asymptotic manifolds which must fold and stretch more and more when approaching the periodic point, Poincaré writes (one should rather say "exclaims"):

> Que l'on cherche à se représenter la figure formée par ces deux courbes et leurs intersections en nombre infini dont chacune correspond à une solution doublement asymptotique, ces intersections forment une sorte de treillis, de tissu, de réseau à mailles infiniment serrées ; chacune des deux courbes ne doit jamais se recouper elle-même, mais elle doit se

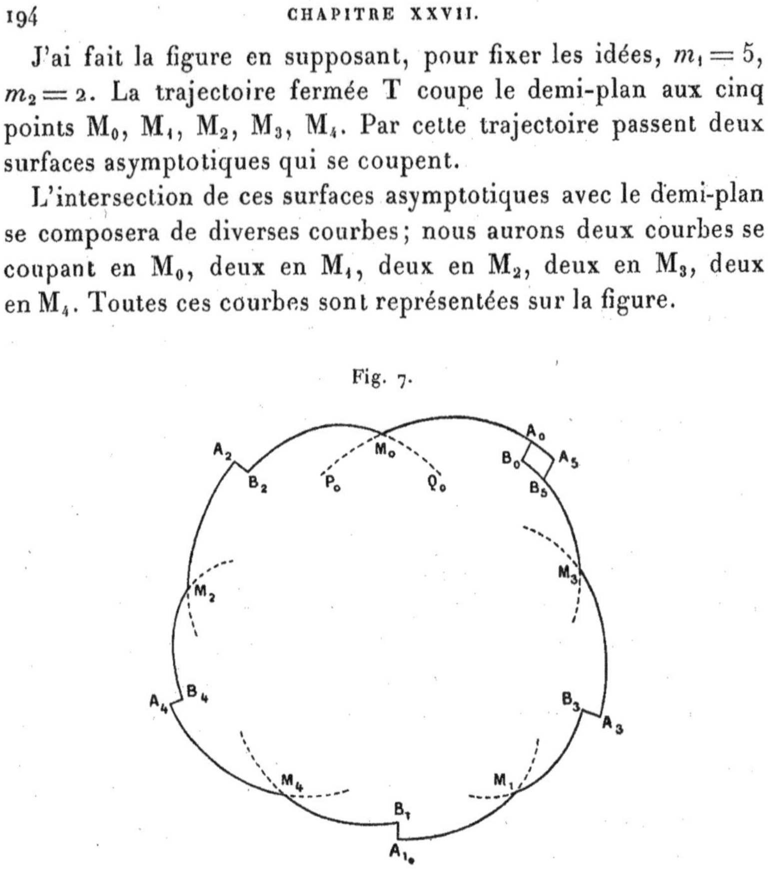

FIGURE 17. The intersection property (this situation cannot occur).

replier sur elle-même d'une manière très complexe pour venir recouper une infinité de fois toutes les mailles du réseau.

On sera frappé de la complexité de cette figure, que je ne cherche même pas à tracer. Rien n'est plus propre à nous donner un idée de la complication du problème des trois corps et en général de tous les problèmes de Dynamique où il n'y a pas d'intégrale uniforme et où les série de Bohlin sont divergentes.[128] (Chapitre XXXIII, Section 397.)

[128]Let us try to represent the figure formed by these two curves and their intersections in infinite number, each corresponding to a doubly asymptotic solution, these intersections form a kind of mesh, of fabric, of infinitely tight network; each of the two curves must never intersect itself, but it must fold back on itself in a very complex way in order to cross an infinite number of times all the meshes of the network.

At the origin of that what some have called (in my opinion, quite inappropriately) "chaos theory", the complexity of the lattice of *homoclinic* or *heteroclinic* intersections of stable and unstable manifolds was partly analysed by Smale (see [Sm1, Sm2, C0]), thanks to the methods of *symbolic dynamics* developed by Hadamard, G.D. Birkhoff, M. Morse, G. Hedlund, ... But, even if the computers give a refined picture (as in Figure 18), understanding all the details remains difficult.

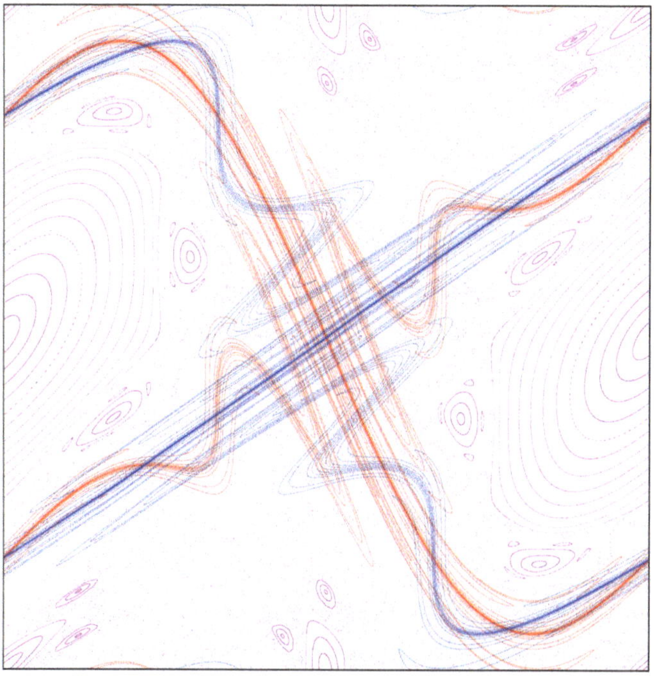

FIGURE 18. A tangle computed by and reproduced with permission of © Carles Simó 2012.

Now comes a question: what is the relation between the divergence of Lindstedt series and the one of Bohlin series? Does the abundance of Fourier coefficients of high order imply both divergences, which would make the link, emphasized by Ramis in [Ra1], between divergence, non-integrability and ambiguity stronger? The proof of non-integrability given in volume I and the proof of the divergence of Lindstedt series which is a consequence, are based on the breaking of periodic tori which gives rise to hyperbolic (i.e., with non-zero exponents) periodic orbits. The splitting of separatrices, linked to the divergence of the Bohlin series, plays no

One will be struck by the complexity of this figure, which I do not even try to draw. Nothing is more likely to give us an idea of the complexity of the Three-Body Problem and in general of all the problems of dynamics where there is no uniform integral and where the Bohlin series are divergent.

role in this proof. Would it be possible that in a system similar to the Three-Body system but more degenerate, the analogue of the Lindstedt series would diverge while the analogue of the Bohlin series would converge? Looking at the complex singularities of the equations of separatrices this looks unlikely but how to prove it? A way to reformulate the question is asking whether Poincaré could have deduced the divergence of the Bohlin series from the divergence of the Lindstedt series.

Finally, the error of the first version of the Memoir was extraordinarily fruitful. Phragmen's questions led Poincaré to discover one of the prominent features not only of the non-integrable Hamiltonian systems, but also more generally of many dissipative generic systems. In his study of the way Poincaré was dealing with divergent series, J.P. Ramis [Ra1], besides explaining how much of the present studies on resummation of divergent series was foreseen by Poincaré, rightly insists on the fact that the discovery of the homoclinic tangle due to the splitting of the separatrices turns the divergence of the series into a positive phenomenon generating a whole geometry. For a first approach, in the non conservative case, of such structures, in particular *Smale's horseshoe*[129] and its connection with the game of heads and tails, that is in mathematical terms with the *Bernouilli shift* and *symbolic dynamics*, I refer to [C0]. For an example of analysis of symbolic dynamics type in a Three-Body Problem (namely the *Sitnikov problem*) see the beautiful book by Jürgen Moser [M5]. For a survey of some symbolic dynamics analysis of complicated motions in the planar Three-Body Problem associated with homoclinic and heteroclinic phenomena near triple collision, see the not less beautiful paper by Rick Moeckel [Moe1]. Finally, it may seem appropriate to conclude this part with a sentence by René Thom in [Th] page 132:

> Ce qui limite le vrai, ce n'est pas le faux, c'est l'insignifiant.[130]

11. Quasi-periodic solutions 2) Analytic aspects: K.A.M. stability

11.1. New methods, Chapter XIII, Section 149

It is in the famous Section 149 of Chapter XIII that Poincaré addresses the issue of convergence, distinguishing the series "with variable frequencies"from the ones with "fixed frequencies". For the first ones, the divergence originates from the "generic" behaviour of the periodic solutions (a priori, but not a posteriori, independently of the non-integrability), About the second ones, he writes:

---

[129]The horseshoe was not discovered at first in the homoclinic tangle. Smale explains that he discovered this paradigmatic model while trying to understand geometrically a paper by Cartwright and Littlewood which showed that a dynamical system could have persistently an infinite number of periodic orbits. This refuted the conjecture he had made that generically a dynamical system would have only a finite number of such orbits. Note that if true, this would have been a complete contrast with the conservative case and the dream of Poincaré that maybe periodic orbits could be generically dense in the phase space.

[130]That which limits the true is not the false, it is the insignificant.

> Il nous reste à traiter la deuxième question ; on peut encore, en effet, se demander si ces séries ne pourraient pas converger pour les petites valeurs de μ, quand on attribue aux x_i^0 certaines valeurs convenablement choisies ... Supposons, pour simplifier, qu'il y ait deux degrés de liberté ; les séries ne pourraient-elles pas, par exemple, converger quand x_1^0 et x_2^0 ont été choisis de telle sorte que le rapport $\frac{n_1}{n_2}$ soit incommensurable, et que son carré soit au contraire commensurable (ou quand le rapport $\frac{n_1}{n_2}$ est assujetti à une autre condition analogue à celle que je viens d'énoncer un peu au hasard) ?
>
> Les raisonnements de ce Chapitre ne me permettent pas d'affirmer que ce fait ne se présentera pas. Tout ce qu'il m'est permis de dire, c'est qu'il est fort invraisemblable.[131]

The existence of solutions of equations (1) which are quasi-periodic with frequencies satisfying Diophantine conditions was announced[132] for the first time by A.N. Kolmogorov in 1954 [K] in the non-degenerate case where the unperturbed Hamiltonian depends effectively on all the actions. Kolmogorov's proof was written by V.I. Arnold [A3] and extended by him to degenerate cases of the type of the planar Three-Body Problem [A1]; a different proof, suitable to the differentiable (i.e., non analytic) case was given by J. Moser [M2], hence the acronym "K.A.M. theory" (see [AKN]); Finally, the convergence of Lindstedt series with Diophantine fixed frequencies,[133] which follows if one knows the analytic dependence of these solutions with respect to the parameter μ, was proved by Moser in [M2]. The invariant tori have a Whitney C^∞ dependence on the frequencies [Laz, Pos]; asked by Kolmogorov, their monogenic dependence on a complex domain containing the Diophantine frequencies has been proved in the particular case of the *Standard map* (see [CMS]).

Kolmogorov's theorem or Moser's *invariant curve theorem* apply directly to the Planar Circular Restricted Three Body Problem to prove the existence of a Cantor set of two-dimensional invariant Lagrangian tori in each three-dimensional fixed energy submanifold or equivalently a Cantor set of invariant curves of the Poincaré return map in a surface of section as defined in Section 9.3. For the (non restricted) Three-Body Problem, the degeneracy of the Kepler Problem seriously complicates the situation.

---

[131] It remains to deal with the second question; we can still, in fact, ask whether the series can not converge for small values of μ, when the x_i^0 are given certain suitably chosen values ... Suppose, for simplicity, that there are two degrees of freedom; could, for example, the series converge when x_1^0 and x_2^0 have been chosen so that the ratio $\frac{n_1}{n_2}$ is non commensurable, and its square is on the contrary commensurable (or when the ratio $\frac{n_1}{n_2}$ is subject to a condition similar to the one I have just stated somewhat at random? The arguments of this chapter do not allow me to affirm that such a case will not arise. All I can say is that it is highly unlikely.

[132] and, according to Sinai, fully proved in his seminar.

[133] Recall (see 5.6) that such a convergence is possible only in the non-degenerate case.

11.2. Arnold's theorem for the planar problem

Arnold's extension [A1] in 1963 of Kolmogorov's theorem was a major breakthrough[134] in the Three-Body Problem, even if applying only to microscopic values of the planetary masses: for the first time, one had a rigorous proof of the existence in the phase space of the planar Three-Body Problem of a set of positive measure of quasi-periodic motions similar to the ones existing in the completely integrable secular system near its singularity, namely, motions of the planets along almost circular and almost coplanar ellipses slowly precessing along the centuries and whose semi-major axes, eccentricities and inclinations remain close to their initial values.

The strategy is close to Poincaré's strategy for constructing Lindstedt series: using the fact that far from mean motion resonances, the averaged system provides a good approximation to the actual system, Arnold reasons as follows: consider the completely integrable system obtained by averaging the fast angles and truncating at fourth order the Birkhoff normal form[135] of the secular part at the singularity corresponding to circular and horizontal motions. The Hamiltonian, defined on a subset of $\mathbb{R}^2 \times \mathbb{T}^2 \times \mathbb{C}^2$ (coordinates $\tilde{\Lambda}, \tilde{\lambda}, \tilde{u} = (\tilde{u}_1, \tilde{u}_2)$), is of the form

$$H = H_0(\tilde{\Lambda}) + \mu H_1(\tilde{\Lambda}, \tilde{u}),$$

$$H_1 = \sum_{1\leq j\leq 2} A_j(\tilde{\Lambda})|\tilde{u}_j|^2 + \sum_{j,k=1}^{2} B_{jk}(\tilde{\Lambda})|\tilde{u}_j|^2|\tilde{u}_k|^2.$$

The fast actions $\tilde{\Lambda}$ and the slow ones $|\tilde{u}_j|^2$ are first integrals of this system and fixing them defines four-dimensional Lagrangian invariant tori in the eight-dimensional phase space. Recall (see 5.6) that the restriction of the flow of the averaged system to the subset defined by the condition that $\tilde{\Lambda}$ avoids a neighborhood of the *secular set* $\mathcal{S}$, can be embedded in the phase space of the planetary problem so as to approximate the actual flow. Provided the frequencies of the invariant tori of this integrable approximation depend in a sufficiently non-degenerate way on the actions, that is if

$$\det \frac{\partial^2 H_0}{\partial \tilde{\Lambda}^2}(\tilde{\Lambda}) \neq 0 \quad \text{and} \quad \det B_{jk} \neq 0,$$

(one then says that there is *torsion*) one can apply refined KAM techniques to prove the persistence – if μ, that is the planetary masses, is small enough and we stay close enough[136] to the secular singularity, i.e., if the eccentricities and inclinations are small enough – of "sufficiently non resonant" invariant tori of the averaged system. The first determinant is easily seen to be different from zero. For

[134] It was also a remarkable introduction to the theory of perturbations using "the frightening formal apparatus of dynamics" and even containing some exercises.

[135] which exists provided one avoids a finite number of secular resonances; Arnold used the Birkhoff normal form up to order six but this is not necessary (see [CP]).

[136] but not too close, unless we center the Poincaré coordinates on a periodic solution of the first sort as explained in 5.6.

the second one, Arnold checks this in the limit of a vanishing ratio of the semi-major axes and argues that analyticity grants the non vanishing of the determinant except for isolated values of this ratio. In this way, he proves the existence of quasi-periodic solutions with 4 frequencies: the two mean motions which are of order $O(1)$ and the secular frequencies of the two perihelia which are of order $O(\mu)$.

11.3. Dealing with the spatial problem

We have seen in Section 5.7 that two resonances do complicate the spatial Three-Body Problem which fortunately disappear both after complete reduction of the rotational symmetry. Arnold, well aware of the first one, $A_4 = 0$, proposed indeed to perform Jacobi's reduction (called *elimination of the node* because it amounts to eliminating the vertical variables) but he did not check himself accurately the presence of torsion. This was done for the first time in 1995 by Philippe Robutel [Rob], using the dedicated computer algebra system TRIP developed by Jacques Laskar [LasRob] to compute the torsion, and also showing theoretically that it is not zero in the limit of an infinitesimal ratio of the semi-major axes.

For an arbitrary number of planets in $\mathbb{R}^3$, two strategies were used in order to rigorously prove Arnold's claim in spite of the resonances (see [Chi, Fe2, Fe3] for surveys):

1) *Herman and Féjoz's proof*, which Herman explained in detail in his seminar during the late nineties. Due to his untimely death he could not write it fully and the proof was completed and written by Jacques Féjoz [Fe4]. The main novelty was to use only the quadratic part of the secular system (i.e., the linearized system) joint with a trick, due to Poincaré and amounting to a reduction, which was to add to the Hamiltonian a small perturbation commuting with it, for example a perturbation proportional to a component of the angular momentum, in order to ensure with a minimum of computations a weak form of non degeneracy[137]. The existence of invariant tori followed from a beautiful normal form theorem,[138] subsuming the degenerate and non degenerate cases as well as the case of invariant tori of non maximal dimension. Finally, an argument of Lagrangian intersection ensured that ergodic[139] invariant KAM tori of the modified Hamiltonian were still invariant by the original one (but not necessarily ergodic any more: one could just assert that they were ergodic after full reduction).

2) *Chierchia and Pinzari's proof* [CP] which is in a sense closer to Arnold's one. Using a modification of Delaunay coordinates due to Deprit, they notice, following [MRL], that both resonances do not impede the construction of a Birkhoff

[137]namely, the *non planarity* of the frequency map, a condition introduced by Arnold and studied by Pyartli, which is sufficient to ensure that the set of actions whose frequency is Diophantine has a positive Lebesgue measure. In agreement with Arnold's law that, in general, a theorem or a notion is not due to the person whose name it bears, this condition is usually called the *Rüsmann condition.*

[138]which was in fact already present in Moser's 1967 paper [M2] in the more general setting of non necessarily conservative systems.

[139]More precisely "bearing linear flows with dense orbits".

normal form after the *partial reduction* consisting in the fixation of the direction of the angular momentum, that is restricting the $(1+n)$-body Hamiltonian to an invariant symplectic submanifold of codimension two (and hence dimension $6n-2$) of the phase space. Restricting to such a submanifold is the natural way to reduce the non commutative symmetry group $SO(3)$ to one of its maximal tori $SO(2)$, and hence *coming back in some sense to the situation of the planar problem.* And indeed, the authors show the existence of torsion in the Birkhoff normal form after such a partial reduction. The Arnold KAM mechanism (more accurately a refined version) then gives ergodic invariant KAM tori of dimension $3n-1$.

To summarize, for the spatial Three-Body Problem, one has the existence of five-dimensional ergodic KAM tori. The five frequencies are easily identified, coming from the mean motions, the precessions of both perihelia and the precession of the common node in the Laplace plane (see 2.3).
Applying the rotation group, that is rotating the direction of the angular momentum, one finds invariant submanifolds of the full phase space which are diffeomorphic to the product $S^2 \times \mathbb{T}^5$ of a sphere by a torus, foliated by the five-dimensional invariant tori: in other words, *the non-commutative part of the symmetry group acts in a trivial way at the dynamical level.* This is not the case of the commutative part: if one reduces completely the rotational symmetry (i.e., if one goes to the quotient by the remaining $SO(2)$-symmetry), one shall find ergodic invariant KAM tori of dimension four in the reduced phase space (as in Herman–Féjoz's proof). When lifted to one of the partially reduced submanifolds, one finds again five-dimensional tori but such tori may be either ergodic or present a resonance between the reduced flow and the rotation.[140]

12. Stroboscopy 2) What we understand of the dynamics of the return map

12.1. Bifurcations: subharmonics

All of Poincaré's studies of periodic solutions show a remarkable insight which prefigure the *theory of singularities* and the studies of *bifurcations.* He proceeds in general from the "general case" to more and more degenerate ones[141] and he does not hesitate mixing quantitative and qualitative tools. For example, in order to be sure that the solutions exist he uses the fact that a zero of the derivative at a maximum point of a holomorphic function is necessarily of odd order, relying for that on a theorem proved at the beginning of his thesis which is exactly the Weierstrass

[140] In the same way, ergodic KAM tori of dimension three in the reduced phase space of the planar Three-Body Problem (the three frequencies being the two mean motions and the frequency of the difference of perihelia) give rise in the non reduced phase space to four-dimensional invariant tori which could be resonant.

[141] The thesis of Anne Robadey [Ro] contains a beautiful study of the notion of "general case" in Poincaré's works on dynamical systems.

preparation theorem.[142] He then can apply the result of an earlier paragraph asserting that a zero of odd order cannot disappear under small perturbation. Here is a typical sentence, in Section 37:

> Une solution périodique ne peut donc disparaître qu'après s'être confondue avec une autre solution périodique.
> En d'autres termes, *les solutions périodiques disparaissent par couples à la façon des racines réelles des équations algébriques.*[143]

The solutions of the second kind (*second genre*, which are born from those of the first kind (themselves divided into the three *sorts* studied in 4.2), are called today *subharmonics*. Their study is the object of Chapters XXVIII, XXX and XXXI. Poincaré describes them as follows in the analysis of his works:

> Nous avons encore les solutions périodiques du deuxième genre; si l'on fait varier de manière continue un des paramètres dont dépend le problème, par exemple l'une des masses, on voit une solution périodique du premier genre se déformer d'une façon continue, sa période restant égale à T. A un certain moment, cette solution se dédouble pour ainsi dire, ou plutôt se détriple, je veux dire qu'à un certain moment on a trois solutions périodiques très peu différentes; l'une d'elles a encore pour période T, les deux autres ont pour période un multiple de T. Ce sont les solutions périodiques du deuxième genre (Analyse des travaux).[144]

In the case of a two degrees of freedom system, taking a surface of section transverse to the periodic solution of period T under study reduces the problem to the existence of periodic points of an area preserving map in the neighborhood of an elliptic fixed point. In Chapter XXX, the study of these solutions is clearly presented as a companion to the study of asymptotic solutions, in today's vocabulary, it is the study of the dynamics attached to an *elliptic* periodic solution while the asymptotic solutions were attached to a *hyperbolic* one (see Figure 19, in a surface of section).

> Dans l'étude des solutions asymptotiques, nous avons supposé que les α_k étaient réels et nous avons annulé une des constantes A sur deux.

---

[142] Weierstrass has written that he had taught this theorem repeatedly in his courses since 1860 but he published it in 1886. Poincaré defended his thesis in 1879.

[143] Therefore, a periodic solution can disappear only after merging with another periodic solution. In other terms, *the periodic solutions disappear by couples as do the real roots of algebraic equations.*

[144] We have still the periodic solutions of the second kind; if we vary continuously one of the parameters of which depends the problem, for example one of the masses, we see a periodic solution of the first kind deform in a continuous way, the period remaining equal to T. At some point, this solution splits into two, or rather three solutions, I mean that at some point we have three periodic solutions which differ only slightly; one of them still has period T, the period of the other two is a multiple of T. Those are the periodic solutions of the second kind (Analysis of the works).

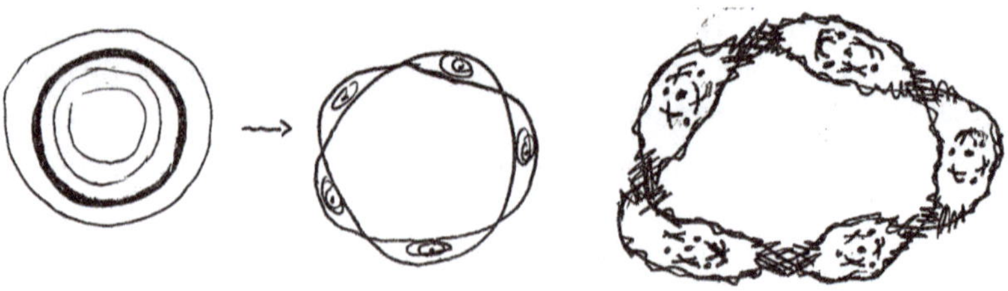

FIGURE 19. $3T$-subharmonics around a periodic solution of the second sort.

> Pour appliquer ce même résultat à l'étude des solutions périodiques du second genre, nous supposerons au contraire que les exposants α_k sont purement imaginaires.[145]

Introducing a parameter μ which controls the exponents of the periodic solution (in the surface of section, a fixed point which one can suppose to be always at the origin), he shows that kT-subharmonics can bifurcate from the T-periodic solution, only in case one exponent of this solution is a multiple of $\frac{2i\pi}{kT}$ (This happens for $\mu = \mu_0$ in Figure 20, drawn in the product of the surface of section and the parameter line).

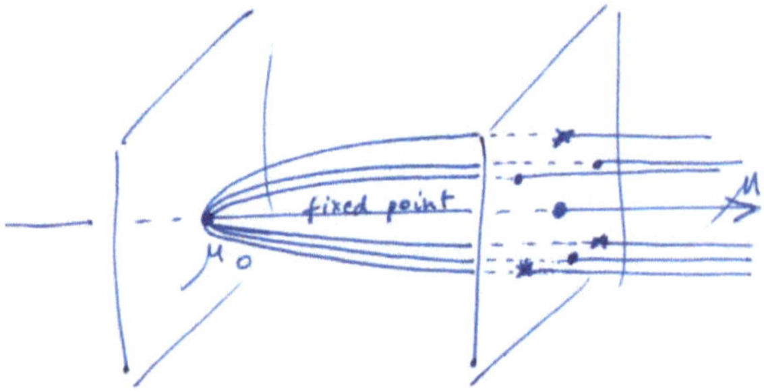

FIGURE 20. The birth of a pair of subharmonics of period $3T$.

Moreover, he proves that

> *il disparaît toujours autant de solutions stables que de solutions instables.*[146]

and, on the other hand, that

[145] In the study of asymptotic solutions, we have supposed that the α_k were real and we have set half of the constants A equal to zero. In order to apply this result to the study of the second kind periodic solutions, we shall suppose on the contrary that the exponents α_k are purely imaginary.
[146] *It always disappears as many stable solutions than unstable solutions.*

> Donc *si, pour une certaine valeur de μ, une solution périodique perd la stabilité ou l'acquiert* (et cela de telle façon que l'exposant α soit nul) *c'est qu'elle se sera confondue avec une autre solution périodique*, avec laquelle elle aura échangé sa stabilité.[147]

The very existence of the subharmonic solutions is reduced to the existence of critical points of a *generating function*: as the flow is symplectic, if we fix T and call respectively (ξ_i, η_i) and (X_i, Y_i) the values at t=0 and $t = T$ of $\big(x_i(t), y_i(t)\big)$,

$$dS = \sum((X_i - \xi_i)d(Y_i + \eta_i) - (Y_i - \eta_i)d(X_i + \xi_i)$$

is an exact differential. Indeed, being somewhat anachronic, we write

$$d^2S = 2(dX_i \wedge dY_i - d\xi_i \wedge d\eta_i) = 0.$$

One concludes if S may be considered as a function of the $X_i + \xi_i$ and $Y_i + \eta_i$. After a chapter (XXIX, see 13.1) dedicated to the *principle of least action* (see Section 13), Poincaré comes back to the subharmonics:

> Je ne puis passer sous silence les rapports entre la théorie des solutions du deuxième genre et le principe de moindre action; et c'est même à cause de ces rapports que j'ai écrit le chapitre XXIX[148] (see 13.1).

He analyses the way the subharmonics wrap around the T-periodic solution and identifies subharmonics with extrema of the action. The *rotation number* is not far.

But let us come back to the (Planar Circular Restricted) Three-Body Problem and more precisely to Section 381 where Poincaré analyses a paper published in 1897 by George Darwin[149] in the *Acta Mathematica* which follows numerically a family of periodic solutions of a Restricted Problem (the Sun, an enormous Jupiter and a small body): as he had done with the continuation of Hill's orbit (see 4.2), but arguing on the necessity of an exchange of stabilities, and pushing at the limit the examination of the different possibilities, he concludes

> Donc, je conclus que les satellites A instables ne sont pas la continuation analytique des satellites A stables. Mais alors que sont devenus les satellites A stables ?
>
> Sur ce point, je ne puis faire que des hypothèses et, pour pouvoir faire autre chose, il faudrait reprendre les quadratures mécaniques de M. Darwin. Mais si l'on examine l'allure des courbes, il semble qu'à un certain moment l'orbite du satellite A a dû passer par Jupiter et qu'ensuite il est devenu ce que M. Darwin appelle un *satellite oscillant*.[150]

---

[147] Therefore, *if for some value of μ, a periodic solution looses or acquires stability* (and this in such a way that the exponent α be equal to zero) *it is because it will have merged with another periodic solution* with which it will have exchanged stability.

[148] I cannot omit to mention the relations between the theory of second kind solutions and the principle of least action; it is even because of these relations that I wrote Chapter XXIX.

[149] A well-known astronomer, the son of sir Charles Darwin.

[150] Hence I conclude that the unstable satellites A are not the continuation of stable satellites A. But then what happened with the stable satellites A? About this, I can only make hypotheses

This is of course a global problem for which there is a lack of tools, but not completely, as the next section shows.

12.2. The last geometric theorem and twist diffeomorphisms of the annulus

> J'ai démontré il y a longtemps déjà, l'existence des solutions périodiques du problème des trois corps; le résultat laissait cependant encore à désirer; car si l'existence de chaque sorte de solution était établie pour les petites valeurs des masses, on ne voyait pas ce qui devait arriver pour des valeurs plus grandes, quelles étaient celles de ces solutions qui subsistaient et dans quel ordre elles disparaissaient.[151]

Poincaré published in 1912, the year of his death, an incomplete proof of what is called today the *Poincaré–Birkhoff fixed point theorem* [P15]. In a both lucid and moving introduction, he explains that he struggled during two years but could not reach a complete proof. Nevertheless, having proved the theorem in all the special cases he had examined and due to its paramount importance, he decided to submit the problem to geometers.

In 1905, he had already addressed more globally the question of periodic solutions on a problem similar to the Restricted Three-Body Problem but "stripped from the secondary difficulties", namely the geodesics on a convex surface [P14, Ro]. Comparing this question with the one completely understood by J. Hadamard [Ha] in the case of negative curvature,[152] he writes:

> J'ai donc abordé l'étude des lignes géodésiques des surfaces convexes; malheureusement, le problème est beaucoup plus difficile que celui qui a été résolu par M. Hadamard. J'ai donc dû me borner à quelques résultats partiels, relatifs surtout aux géodésiques fermées, qui jouent ici le rôle des solutions périodiques du problème des trois corps.[153]

The fixed point theorem will be proved by Birkhoff during the year following Poincaré's death [Bi1] (for a radically different proof, see [BH]); it asserts the existence of at least two fixed points of an area preserving *twist diffeomorphism*[154]

---

and, in order to get something else one should continue the mechanical quadratures of M. Darwin. But, if one examines the shape of the curves, it seems that at some moment the orbit of the satellite must have passed near Jupiter and that after that he became what M. Darwin calls an *oscillating satellite.*

[151] I showed long ago, the existence of periodic solutions of the Three-Body Problem; nevertheless, the result was still unsatisfactory; for, if the existence of each type of solution was established for small values of the masses, one did not see what would happen for larger values, which ones of these solutions remained and in what order they disappeared.

[152] in which case, the periodic geodesics are indeed dense in the phase space and moreover a full description of the flow with *symbolic dynamics* is available.

[153] So, I approached the study of geodesic lines of convex surfaces; unfortunately the problem is much more difficult than the one solved by M. Hadamard. Hence I had to content myself with some partial results, mainly concerning closed geodesics, which play here the role of periodic solutions of the Three-Body Problem.

[154] i.e., a diffeomorphism of $S^1 \times [0,1]$ which admits a lift to a diffeomorphism of $\mathbb{R} \times [0,1]$ which is increasing on one boundary and decreasing on the other.

of the annulus $S^1 \times [0, 1]$. The application to the Restricted Three-Body Problem with a large Jacobi constant, comes from the fact that, the return map in the Birkhoff annulus of section[155] is a *monotone twist map*,[156] which allows application of the theorem to its iterates in order to get periodic orbits. But in fact, the full force of the theorem is not needed in this case or in the companion case of an area preserving diffeomorphism of the plane in the neighborhood of a generic elliptic fixed point. The main features which are understood are represented on Figure 21 (see [Ze, C9, Le]):

1) The *KAM invariant curves*[157] on which the return map is analytically conjugate to a rotation of angle $2\alpha\pi$ with α an irrational satisfying a *Diophantine condition.* They imply STABILITY! Generically, the union of these invariant curves is a set whose transversal structure is Cantor-like. The annuli in between two invariant curves (and not containing in their interior any invariant curve homotopic to the boundary) are called *Birkhoff zones of instability.* The dynamics in such a zone is quite complicated and certainly not completely understood. The main features are

2) "well-ordered" periodic orbits,[158] obtained by minimization of a generating function; the homoclinic and heteroclinic tangles associated with the invariant manifolds of the hyperbolic periodic orbits (see [Ze]);

3) the Aubry–Mather invariant Cantor sets and the invariant manifolds attached to them, part of which make the *weak KAM solutions.* These sets, whose existence was initially obtained by S. Aubry and J. Mather, in each case by minimizing the action functional, were shown by A. Katok to be limits of periodic orbits, thanks to the Birkhoff a priori Lipschitz estimates resulting from the monotone twist condition;

4) diffusing orbits found by Birkhoff and refined by Mather using a variational approach, which go from one component of the boundary to the other one and may be imposed to pass through the Aubry–Mather sets in a prescribed way;[159]

5) and the same phenomena repeated around the subharmonics, etc. But we still do not know whether periodic points are dense or if hyperbolic behaviour (that is non-zero Lyapunov exponents) exist on a set of positive Lebesgue measure.

The Poincaré–Birkhoff fixed point theorem, which makes more global part of the above results, has played a very important role in the development of *symplectic topology,* in particular through the proofs of various instances of the *Arnold conjecture* relating the number of fixed points of Hamiltonian diffeomorphisms of

[155] See Figure 15 and [Co].

[156] i.e., there exists an identification of the annulus of section to $S^1 \times [0, 1]$ such that the image by the return map of any "vertical" segment $\theta_0 \times [0, 1]$ is a graph over a part of the circle (see [C9, Le]).

[157] which, in the case of a return map, are the trace of the KAM tori on a surface of section.

[158] i.e., circularly ordered as a periodic rotation; thanks to the monotone twist condition, one gets periodic orbits much better controlled than the ones given by the Poincaré–Birkhoff fixed point theorem applied to iterates of the return map.

[159] The variational techniques having led to these results are currently used in order to prove the generic existence of diffusion in 2.5 or 3 degrees of freedom Hamiltonian systems.

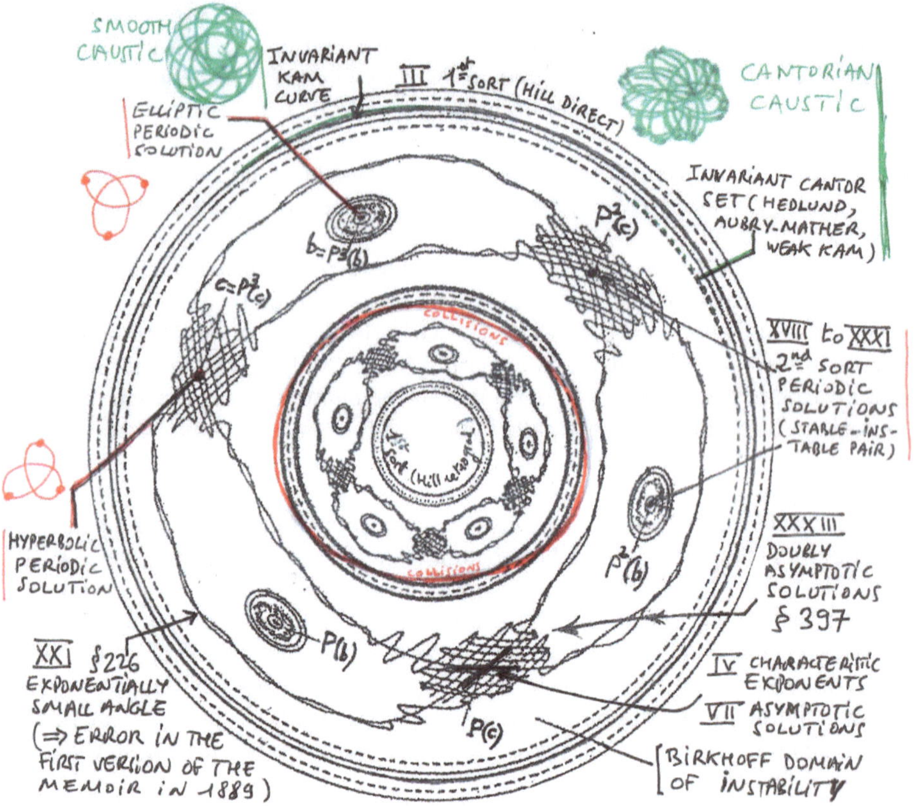

FIGURE 21. The Planar Circular Restricted Three-Body Problem in the case of a high Jacobi constant (Birkhoff global annulus of section). Figure 21 is a revised version of Figure 39 in [C0], published with kind permission of ©Encyclopedia Universalis France S.A. 2005. All Rights Reserved.

compact manifolds and Morse theory. The breakthrough was the paper [CZ] by Charles Conley and Eddy Zehnder, followed by a reduction to the finite dimension of their proof by Marc Chaperon [Chap], well in the spirit of Poincaré's use of generating functions. Describing this would lead us too far astray, hence I refer to [Au2] for this part of the story and the references.

12.3. Computer experiments, assisted proofs and the geometrization of phase space

In the twentieth century, an extensive search for families of periodic solutions in the Restricted Three-Body Problem was accomplished, first by mechanical quadratures at the Copenhagen Observatory (Stromgren), later using computers by Hénon at the Nice Observatory, Broucke, and others. The books by Szebehely

[Sz], Hénon [He], Bruno [Br] describe the Planar Circular Restricted Three-Body Problem both theoretically and numerically.

Particularly interesting for mission design are the Halo orbits in the spatial Restricted Problem, which bifurcate from a planar Lyapunov family originating from a collinear relative equilibrium. Other much studied special cases are the collinear problem with the remarkable periodic (regularized) solution discovered by Schubart (see an animation in [C4]) and the isosceles problem, where one body moves on a line, while the two others, with the same mass, move symmetrically on the orthogonal line (resp. plane in the spatial case).

Today, the power of computers allows us to get a pretty good understanding of autonomous 2 degrees of freedom systems like the Restricted Problem or its simplified version called the *Hill problem*. For example, an extensive study of the phase space of the related Hill's problem was completed by Simó and Stuchi [ST]; computer assisted proofs of the existence of KAM tori (and hence of stability) for realistic (Sun-Jupiter-Ceres or Sun-Jupiter-Victoria), but truncated to a trigonometric polynomial, Restricted Three-Body Problems, were given by Alessandra Celletti and Luigi Chierchia in [CC1, CC2]. On the other hand, [GaK, GaU], among others, provide computer assisted proofs of the existence of orbits diffusing in a Birkhoff zone of instability for the Planar Circular Restricted Three-Body Problem.

Finally, results of C. Simó and his collaborators show the important role played by invariant manifolds or, more generally, of center manifolds of (families of) periodic solutions, in determining the domains of *practical stability of celestial bodies.* See for instance [Si, SST]. In some sense, this shows that Poincaré was not completely wrong when trying to prove stability from the behavior of these invariant manifolds.

13. A great principle of physics and some collisions

In *The value of Science* [P13], Poincaré puts the *Principle of least action* on the same footing as the great conservation principles (energy, mass, action-reaction), the principle of degradation of energy and the relativity principle. Due to its global character it may appear at first sight closer to theology than to physics and indeed, in Chapter VIII of *Science and Hypothesis* Poincaré wonders:

> L'énoncé même du principe de moindre action a quelque chose de choquant pour l'esprit. Pour se rendre d'un point à un autre, une molécule matérielle soustraite à l'action de toute force mais assujettie à se mouvoir sur une surface, prendra la ligne géodésique, c'est-à-dire le chemin le plus court.
>
> Cette molécule semble connaître le point où on veut la mener, prévoir le temps qu'elle mettra à l'atteindre en suivant tel et tel chemin, et choisir ensuite le chemin le plus convenable. L'énoncé nous la présente pour ainsi dire comme un être animé et libre. Il est clair qu'il vaudrait

> mieux le remplacer par un énoncé moins choquant, et où, comme diraient les philosophes, les causes finales ne sembleraient pas se substituer aux causes efficientes.[160]

In [Fey] (*volume I, Chap. 26, par. 5: A more precise statement of Fermat's principle*), Richard Feynman asks the same question, but with this time the answer given by quantum electrodynamics, that is by the *principle of stationary phase*:

> Instead of saying it is a causal thing, that when we do one thing, something else happens, and so on, it says this: we set up the situation, and *light* decides which is the shortest time, or the extreme one, and chooses that path. But *what* does it do, *how* does it find out? Does it *smell* the nearby paths, and check them against each other? The answer is, yes, it does in a way.

13.1. Least action and instability

Chapter XXIX of *The New Methods* opens with a crash course in elementary *Calculus of variations* applied to the action integrals $\int \sum_i p_i dq_i$ (the *Maupertuisian action*) when fixing the energy and $\int (\sum_i p_i dq_i - H dt)$ (the *Hamiltonian action*) when fixing the time interval.[161] Of course, in the case of the Three-Body Problem, $\sum p_i \frac{dq_i}{dt} - H$ is simply the Lagrangian in the disguise provided by the *Legendre transformation* and I suppose that, if there is a single notion which need not be explained to physicists, it must be the Lagrangian action $\int L dt$.

Then, developing the note [P10] which was based on a simple geometric reasoning, he proves that a periodic solution which locally minimizes the action must be dynamically unstable:

> En résumé, pour qu'une courbe fermée corresponde à une action moindre que toutes les courbes fermées infiniment voisines, il faut et il suffit que cette courbe fermée corresponde à une solution périodique instable de la première catégorie.[162]

In particular, a "generic" closed geodesic on a surface locally minimizes the length among nearby closed loops if and only if it is hyperbolic (and hence unstable), i.e., if its non-trivial exponents are real. For example, on a torus of revolution around the z axis, the intersections of the torus with the horizontal plane of symmetry

[160] The very statement of the principle of least action has something shocking to the mind. To go from one point to another one, a material molecule, taken away from the action of any force, but constrained to move on a surface, will follow the geodesic line, i.e., the shortest path. It seems that this molecule knows the point where one wants it to go, that it anticipates the time needed to reach it along such or such path, and then chooses the most convenient path. In a sense, the statement presents this molecule as a free animated being. It is clear that it would be better to replace it by a less shocking statement where, as philosophers would say, the final causes would not appear to replace the efficient ones.

[161] See Remark 1 at the end of Section 8.1.

[162] In short, for a closed curve to correspond to a lesser action than all closed curves which are infinitely close, it is necessary and sufficient that the closed curve corresponds to an unstable periodic solution of the first category.

are respectively unstable for the short one, along which the curvature is negative, and stable for the long one, along which the curvature is positive: along the first one there are no conjugate points while they exist on the second one.

Such assertions hold only for mechanical systems with two degrees of freedom. Indeed, examples studied by Marie-Claude Arnaud in [Ar] show that in higher dimensions, a locally action minimizing periodic solution may possess only two directions of instability transverse to the flow in its energy level.

13.2. Minimizing the action: the note of 1896

In 1896, Poincaré publishes the short note [P5], titled *On the periodic solutions and the least action principle.* The note of 1897 which was quoted in Section 13.1 will have almost the same title, but this one is much more ambitious, even if not directly applicable to the Newtonian force. Indeed, in the analysis of his works, Poincaré writes:

> Je suis revenu sur ces solutions périodiques et je les ai étudiées en détail. Les procédés dont je me suis servi pour démontrer leur existence sont très simples et se ramènent au calcul des limites.
>
> Mais on peut arriver à cette démonstration par une voie toute différente, qu'il pourra être souvent utile d'adopter, mais dont je n'ai pas encore tiré tout le parti possible. Supposons par exemple que l'on recherche les géodésiques d'une surface indéfinie présentant la même forme générale qu'un hyperboloïde à une nappe. On sera certain alors qu'il doit y avoir une géodésique fermée (correspondant à une solution périodique) parce que, parmi toutes les courbes fermées que l'on peut tracer sur la surface et qui en font le tour, il doit y en avoir une qui est plus courte que toutes les autres.
>
> Les mêmes principes sont susceptibles d'être appliqués à divers problèmes de Mécanique, grâce au principe de moindre action que l'on peut employer soit sous la forme que lui a donnée Hamilton, soit sous celle que lui a donnée Maupertuis. Je n'ai fait qu'esquisser cette méthode dont il y a sans doute encore beaucoup à tirer.[163]

Referring to [C2] for more details and references, I quickly describe what Poincaré does: he is looking for *relative* periodic solutions of the planar Three-Body Problem, i.e., solutions which after the period T are such that the triangle formed by

[163] I came back to these periodic solutions and I studied them in detail. The methods I used to prove their existence are very simple and can be reduced to the calculus of limits. But one can arrive to this proof by a completely different path, which it will often be useful to follow, but which I did not fully exploit. Suppose, for instance, that one is looking for the geodesics of an indefinite surface whose general shape is the one of a one-sheeted hyperboloid. One can be sure that there exists a closed geodesic (corresponding to a periodic solution) because, among all the closed curves going around the surface which one can draw, one must be shorter than the others. Thanks to the Principle of Least Action, used in the form given by Hamilton or in the form given by Maupertuis, the same principles can be applied to various problems of Mechanics. I have only sketched this method from which there is still probably much to get.

the three bodies is the same up to a rigid rotation around the center of mass (which one can suppose to be fixed). Given a relative loop of configurations of the three bodies in the plane (i.e., a family parametrized by the interval $[0,T]$ of triangles changing their shape and position but keeping their center of mass fixed and coming back to the same triangle up to a rotation $\theta \mod 2\pi$), he calls respectively $\theta, \theta+2k\pi, \theta+2l\pi,\ k,l\in\mathbb{Z}$ the total (i.e., not $\mod 2\pi$) angles described by the three sides of the triangle during one period. Then, looking for the simplest possible solutions in this class, he minimizes the Lagrangian action $\int_0^T \left[\frac{1}{2}\sum_i m_i ||\frac{d\vec{r}_i}{dt}||^2 + \sum_{i<j}\frac{m_i m_j}{||\vec{r}_i-\vec{r}_j||}\right] dt$ of the Three-Body Problem in the subspace of those relative loops for which the integers k and l are fixed. In fact this was fixing the homology class of the loop of relative configurations. It is interesting to recall that the year before, Poincaré had defined the fundamental group. Choosing k and l different from zero implies coercivity, i.e., forbids a minimizer to be "at infinity". Actually, it is only around 1930 that the existence of an absolutely continuous minimizer was proved by Tonelli provided coercivity holds, its regularity following from Weierstrass theory; but Poincaré does not even mention the question. He does not mention either the major problem, the possibility for a minimizer of having collisions and hence not being a true solution: it is an important result of Sundman around 1913 that, in a motion of n point masses under Newtonian attraction, if two or more bodies collide at a certain instant t_0, their mutual distances and mutual velocities satisfy the same estimates as do colliding bodies in the Two-Body Problem

$$||\vec{r}_i-\vec{r}_j|| = O(|t-t_0|^{\frac{2}{3}}) \quad \text{and} \quad ||\frac{d\vec{r}_i}{dt}-\frac{d\vec{r}_j}{dt}|| = O(|t-t_0|^{-\frac{1}{3}}).$$

This relative weakness of the Newtonian attraction implies that *the action stays finite even if a collision occurs* and leaves the possibility that minimizers have collisions. Well aware of the problem, Poincaré avoids it by supposing from the start a *strong force attraction*, that is a *force function U* which is a sum of terms of the form $\frac{kmm'}{r^n}$, with *n at least equal to* 2. He may then argue later on:

> Je dis maintenant que, si a et c se rencontrent, l'action est infinie.
>
> En effet, l'action sera du même ordre de grandeur que $\int 2U dt$, ou que $\int 2\sqrt{U} dr$, ou que $2kmm' \int \frac{dr}{r^{\frac{n}{2}}}$, c'est-à-dire infinie si $n \geq 2$. Or, on a $n = 2$ si, comme nous le supposons, l'attraction s'exerce en raison inverse du cube des distances.[164]

This is cheating of course (and also forgetting the possibility of triple collisions), but at the same time very typical of Poincaré's bold exploration of a new domain: going further at a quick pace, while being well aware of the questions to which he should come back ... if time permits.

[164] I say now that , if a and c meet, the action is infinite. Indeed, the action will be of the same order of magnitude as $\int 2U dt$, or as $\int 2\sqrt{U} dr$, or as $2kmm' \int \frac{dr}{r^{\frac{n}{2}}}$, that is infinite if $n \geq 2$. But, one has $n = 2$ if, as we suppose, the attraction is proportional to the inverse cube of the distances.

As we know, there was no time for Poincaré to come back to this question and this note was forgotten during a century. For instance, it is not mentioned by William B. Gordon who, in 1977, shows that minimizers of the action $\int_0^T \left(\frac{1}{2}||\frac{d\vec{r}}{dt}||^2 + \frac{k}{||\vec{r}||}\right) dt$ of the *Kepler Problem* in the plane (Newtonian attraction by a fixed center at the origin of $\mathbb{R}^2$) among loops of index k around the origin, are the Kepler ellipses of period T if $k = \pm 1$ while if $|k| \geq 2$, they lie on the boundary of the chosen set of loops, being the ejection-collision solutions of period T, whose action is strictly smaller than the one of the elliptic solutions of period T/k described k times! Indeed, constraining the homology of the loop of configurations (be it relative or absolute) most of the time leads to minimizers with collisions (there are exceptions found by A. Venturelli, K.C. Chen,...). But, if one replaces the homology constraint by a symmetry constraint, the method becomes remarkably efficient, leading in particular to new classes of periodic or quasi-periodic solutions, the *choreographies* and the *Hip-Hops* (see the references in [C2, Mont1, Mont2],). The important fact behind this success is *Marchal's theorem* [Ma2, C8, C5] which asserts that, for the N-body problem in the plane or in space, minimizers of the action in given time between two fixed configurations have no collision.
For nice examples of a different use of the variational principle, see [GaK, Moe2]

13.3. Solutions of the second species

In spite of the fact that, in Poincaré's sketch of the existence of these solutions, there is no appeal to the principle of least action, a recent proof by Bolotin and Negrini [BN] is based on it, hence the choice to put this section at the end of the one on the principle of least action. Being interested mainly in planetary systems of the type of the solar system, Poincaré was not much concerned with collisions.[165] In the note [P5] that was discussed in the previous section, he even surreptitiously changed the law of attraction in order to avoid them. In *The New Methods*, there is only one chapter, which indeed displays a quite remarkable intuition, where he copes with solutions approaching collisions, in fact only two-body collisions. The idea is simple and beautiful: in the limit of zero masses, the planets follow Keplerian orbits till they have a close encounter (i.e., in the limit, a collision) and then shift to another pair of Keplerian ellipses:
A full symbolic dynamics of such almost collision orbits has been constructed by Bolotin: it implies the existence of solutions with an erratic diffusion of the angular momentum and a much slower one of the Jacobi constant. Existence in the full Three-Body Problem is proved in [BN] where one can find references to the previous works in the Restricted (Circular or Elliptic) Problem.

If one replaces almost double collision by almost triple collision, the situation becomes tremendously richer. The way to analyse it is to compactify by adding the so-called *McGehee collision manifold* (see [McG, C7]). A description of some of

[165]Nevertheless, as we saw in Section 8.1, he understood the scaling invariance of the equations, fundamental in the studies of collisions.

the complexity it implies in the phase portrait of the planar Three-Body Problem can be found in the beautiful survey paper by Rick Moeckel [Moe1].

14. Resonances 3) diffusion

14.1. Arnold diffusion

Kolmogorov's result implies a very strong stability result in the (planar circular) Restricted Problem. Indeed invariant curves are barriers, playing the role that Poincaré had wanted to assign to the separatrices of the hyperbolic periodic orbits in the first version of his Memoir of 1889. Indeed, as noticed by Arnold in the Russian version of Poincaré's works, in the case of 2 degrees of freedom, the existence of KAM invariant tori implies the existence of non trivial C^∞ integrals.

As is well known today, the situation is much more complex in the non restricted case where, the number of degrees of freedom being strictly greater than 2, Lagrangian invariant tori do not separate an energy manifold. This leaves room for a very slow diffusion of the trajectories not belonging to the invariant tori: remaining a long time around the KAM tori which are quite sticky, they could a priori escape very far. The first (highly non generic) example of this phenomenon was given by Arnold in 1964 [A4].

Lower bounds on diffusion time for close to integrable Hamiltonian systems were first given by Nekhoroshev [Ne]. An example of application to the Three-Body Problem is [Nie]. Proving that this phenomenon occurs in a specific (and not *ad hoc*) problem or even generically is very hard; a recent example for the Three-Body Problem is [FGKR] for the Planar Elliptic Restricted Three-Body Problem. For the long-term evolution of the full solar system, see [Las1].

14.2. The oldest open question in dynamical systems

In the 1998 International congress of mathematicians, which took place in Berlin, Michael Herman [Her] ended his beautiful survey of open problems with the following question:

> For the n-body problem in space, we will suppose $n \geq 3$.
>
> • The center of mass is fixed at 0.
>
> • On the energy surface we C^∞-reparametrize the flow by a C^∞ function φ_e (after reduction of the center of mass) such that the flow is complete: we replace H by $\varphi_e(H - e) = H_e$ so that the new flow takes an infinite time to go to collisions (φ_e is a C^∞ function outside collisions).
>
> Following G.D. Birkhoff [Bi] (who only considers the case $n = 3$ and the angular momentum $\neq 0$) (see also A.N. Kolmogorov [K]), we ask:
>
> **Question.** *Is for every e the non-wandering set*[166] *of the Hamiltonian flow of* H_e *on* $H_e^{-1}(0)$ *nowhere dense in* $H_e^{-1}(0)$*?*

[166] A point is wandering if some neighborhood of it, when propagated by the flow of the differential equation, never comes back to intersect itself.

In particular, this would imply that the bounded orbits are nowhere dense and no topological stability occurs.
It follows from the identity of Jacobi–Lagrange that when $e \geq 0$, every point such that its orbit is defined for all times, is wandering.
The only thing known is that, even when $e < 0$, wandering sets do exist (Birkhoff and Chazy, see Alexeyev [Ale]).

The fact that the bounded orbits have positive Lebesgue measure when the masses belong to a non empty open set, is a remarkable result announced by V.I. Arnold [A1] (Arnold gives only a proof for [the] planar 3-body problem and if the author is not mistaken, Arnold's claim is correct).
In some respect Arnold's claim proves that Lagrange and Laplace, against Newton, are correct in the sense of measure theory and that, in the sense of topology, the above question, in some respect, could show Newton to be correct.[167]

In order to know the answer of astronomers to this question, the best is to read the report by Jacques Laskar in this seminar [Las1]. Indeed, as he told me in front of a Chinese tea, for realistic masses of the planets, instability seems likely, even in the sense of measure!

One may contrast this with the words of Poincaré at the beginning of [P8].

> Les personnes qui s'intéressent aux progrès de la mécanique céleste, mais qui ne peuvent les suivre que de loin, doivent éprouver quelque étonnement en voyant combien de fois on a démontré la stabilité du système solaire.
>
> Lagrange l'a établie d'abord, Poisson l'a démontrée de nouveau, d'autres démonstrations sont venues depuis, d'autres viendront encore. Les démonstrations anciennes étaient-elles insuffisantes, ou sont-ce les nouvelles qui sont superflues ?
>
> L'étonnement de ces personnes redoublerait sans doute, si on leur disait qu'un jour peut-être un mathématicien fera voir, par un raisonnement rigoureux, que le système planétaire est instable.
>
> Cela pourra arriver cependant ; il n'y aura là rien de contradictoire, et cependant les démonstrations anciennes conserveront leur valeur.[168]

[167] It is in the *Opticks* [N] that Newton says that God must from time to time set the solar system back into order:

> For while Comets move in very eccentric Orbs in all manner of Positions, blind Fate could never make all the Planets move one and the same way in Orbs concentric, some inconsiderable Irregularities excepted, which may have risen from the mutual Actions of Comets and Planets upon one another, and which will be apt to increase, till this System wants a Reformation.

while, as we saw in 3.3, Lagrange and Laplace show stability at first order by studying the secular system linearized at circular and coplanar motions.

[168] The persons interested in the advances of celestial mechanics but who can only follow them from far, must be somewhat astonished when seeing how many times the stability of the Solar System was proved. Lagrange first proved it, then Poisson, other proofs followed, other are still

In fact, in his conclusion, greatly overestimating the importance of dissipative forces,[169] in particular of the tides, he concludes that their action makes illusory, at the astronomical level, any further progress in the mathematical analysis of the abstract problem of stability.

15. Surprises of a eulogy

On December 15th 1913, Gaston Darboux pronounces at the Paris Academy of Science the eulogy of Henri Poincaré [D]. I do not resist the pleasure of reproducing the beginning of his Section VII:

> Poincaré ne pouvait pas ne pas avoir conscience de la haute valeur de ses écrits ; d'autres auraient réclamé des récompenses, lui ne demandait rien. Nous le regardions tous comme le plus fort d'entre nous. Il n'a jamais cherché à nous devancer. Nul ne pouvait prévoir les vides nombreux que la mort allait faire dans la section de Géométrie. Pour le faire arriver plus vite, pour lui ménager une place dans la section d'Astronomie, on lui signalait les applications que les théories par lui découvertes pouvaient avoir en Mécanique céleste. Il suivait docilement ces indications, s'occupait du problème des trois corps, des figures des corps célestes, et trouvait tout naturel de laisser passer devant lui tous les anciens.[170]

16. A seminar

From 1988 to 1990, Jacques Laskar and the author held in the Bureau des Longitudes (of which Poincaré had been a member since 1893 and three times president, a three-year-long seminar dedicated to reading the three volumes of *The New Methods of Celestial Mechanics*. Being attended by astronomers and mathematicians, the seminar was at the origin of the creation of ASD (*Astronomie et Systèmes Dynamiques*), a group in which we try to perpetuate the collaboration between astronomers and mathematicians. We knew that Poincaré's researches on differential equations, and in particular on the Three-Body Problem, had created whole

---

to come. Were the ancient proofs no sufficient, or is it that the new ones are superfluous? The astonishment of these persons would certainly increase, if they were told that it may happen that some day a mathematician will show, by a rigorous reasoning, that the planetary system is unstable. Nevertheless, this might well happen; and yet there will be no contradiction, and the ancient proofs will keep their value.

[169]Indeed, according to [Las2], in the solar system (but not in some extrasolar systems) the conservative diffusion of trajectories greatly dominates the dissipative forces.

[170]Poincaré could not be unaware of the high value of his writings; others would have claimed awards, he asked nothing. We all considered him as the strongest of us all. He never attempted overtaking us. No one could foresee the many gaps that death would cause in the Geometry section. In order to make him arrive faster, to secure for him a seat in the Astronomy section, we pointed to him the applications that the theories he had discovered could have in Celestial Mechanics. He was obediently following these indications, handling the Three-Body Problem, the figures of celestial bodies, and finding it quite natural to let pass before him the elders.

chapters of today's mathematics: dynamical systems, differential forms, ergodic theory, topology, ..., but we have discovered in the course of these three years the visionary precision with which these three volumes exposed ideas which we had thought recent. The items [S1, S2, S3] were published at that time.

17. Thanks

to Alain Albouy, Jacques Féjoz, Jacques Laskar, Philippe Robutel, and all members of the group ASD (Astronomy and Dynamical Systems) for years of illuminating discussions on these matters and others. Thanks to Daniel Bennequin, Alexey Borisov, Thierry Combot, Thibault Damour, Sylvio Ferraz-Mello, Yanning Fu, Hugo Jiménez-Pérez, Vadim Kaloshin, Yvette Kosmann-Schwarzbach, Christian Marchal, Jessica Massetti, Rick Moeckel, Carles Simó, Jacques-Arthur Weil, Lei Zhao, for precisions, documents, references, corrections. Special thanks to Jacques Féjoz, Bertram E. Schwarzbach, Tadashi Tokieda, Scott Walter, for everything which sounds like good English.

18. Regret

Celui de vivre à une époque où la belle langue de Poincaré est de moins en moins audible, comme en témoigne l'écriture en anglais de ce texte, demandée par les organisateurs du "séminaire Poincaré", que je remercie néanmoins pour leur souhait de rendre à Poincaré l'hommage des physiciens d'aujourd'hui (*sorry, the tribute of contemporary physicists*).

19. Note on the references

While writing this text, I have made free use of parts of my former papers on Poincaré's works on the N-body Problem: [S1, S2, C1, C2, C3]. The last three are available on my page of preprints:
`http://www.imcce.fr/Equipes/ASD/person/chenciner/chenciner.html`
Giving all the sources for the concepts involved was of course beyond the scope of this paper, hence I have often given priority to easier-to-read expositions and in particular to the ones I understand best, that is my own. Nice general descriptions respectively of the Two-body and Three-Body Problem are [Alb2, Mal], as well as the classical reference [Wi] which has very interesting historical notices. Some of the figures are taken from the slides of a lecture given on *The New Methods* at Paris Observatory on July 9 of this year (hence the annotations in French). These slides are available at `http://www.imcce.fr/poincare2012/` or at my address above under the heading "Quelques conférences". Finally, for those who are able to read French, I remind you that many of Poincaré's papers and letters are available, with many other documents, at the address `http://www.univ-nancy2.fr/poincare/documents/bd1.html`.

References

[S1] A. Chenciner, *Intégration du Problème des deux corps par la méthode de Hamilton–Jacobi : les coordonnées action-angle de Delaunay*, J. Laskar *Les variables de Poincaré et le développement de la fonction perturbatrice*, note technique S026 du Bureau des Longitudes, 1989.

[S2] A. Chenciner, *Séries de Lindstedt*, note technique S028 du Bureau des Longitudes, 1990.

[S3] S. Ferraz-Mello, *The Method of Delaunay*; A. Jupp *Chapter XIX Bohlin Method*, note technique S036 du Bureau des Longitudes, 1992.

[AA] K. Abdullah et A. Albouy, *On a strange resonance noticed by M. Herman*, Regular and chaotic dynamics **6**, 4, 421–432 (2001).

[Alb1] A. Albouy, *Histoire des équations de la mécanique analytique : repères chronologiques et difficultés*, in "Siméon-Denis Poisson, les mathématiques au service de la science" (Y. Kosmann-Schwarzbach ed.), éditions de l'École polytechnique, to appear.

[Alb2] A. Albouy, *Lectures on the Two-body Problem*, Classical and Celestial Mechanics (The Recife lectures), H. Cabral and C. Diacu eds., Princeton University Press (2002).

[AC] A. Albouy & A. Chenciner, *Le problème des n corps et les distances mutuelles*, Inventiones Mathematicæ**131**, 1, 151–184 (1997).

[Ale] V.M. Alexeyev, *Sur l'allure finale du mouvement dans le problème des trois corps*, ICM Nice **2**, 893–907 (1970), & *Final motions in the three-body problem and symbolic dynamics*, Russian Math. Surveys **36**, 4, 181–200 (1981).

[d'Al] d'Alembert, *Œuvres complètes de Jean Le Rond D'Alembert : premiers textes de mécanique céleste* 1747–1749, ed. by Michelle Chapront-Touzé, CNRS ed.

[Ar] M.C. Arnaud, *On the type of certain periodic orbits minimizing the Lagrangian action*, Nonlinearity **11**, 143–150 (1998).

[A1] V.I. Arnold, *Petits dénominateurs et problème de la stabilité du mouvement en mécanique classique et céleste*, Usp. Mat. Nauk **18**, 91–192 (1963).

[A2] V.I. Arnold, *Mathematical Methods of Classical Mechanics*, Graduate Texts in Mathematics, Vol. 60, 2nd edition, Springer, 1989.

[A3] V.I. Arnold, *Proof of a theorem of A.N. Kolmogorov on the invariance of quasi-periodic motions under small perturbations of the Hamiltonian*, 1963.

[A4] V.I. Arnold, *Instability of dynamical systems with several degrees of freedom*, Dokl. Akad. Nauk SSSR **156**, n° 1, 9–12 (1964) (Russian); English translation: Sov. Math. Dokl. **5**, 581–585 (1964).

[AKN] V.I. Arnold, V.V. Kozlov & A.I. Neishtadt, *Mathematical Aspects of Classical and Celestial Mechanics*, 3rd ed., Dynamical Systems III, Encyclopædia of Mathematical Sciences, vol. 3, Springer 2006.

[Au] M. Audin, *Intégrabilité et non-intégrabilité de systèmes hamiltoniens [d'après S. Ziglin, J. Morales-Ruiz, J.P. Ramis]*, séminaire Bourbaki 884, mars 2001 `http://www.numdam.org/numdam-bin/feuilleter?id=SB_2000-2001__43_`.

[Au2] M. Audin, *Vladimir Igorevitch Arnold and the invention of symplectic topology*, preprint 2012.

[BG] J. Barrow-Green, *Poincaré and the Three-Body Problem* AMS, History of Mathematics, vol. 11, (1997).

[Bi] G.D. Birkhoff, *Dynamical Systems*, Chapter IX, AMS, (1927).

[Bi1] G.D. Birkhoff, *Proof of Poincaré's geometric theorem*, Trans. Amer. Mat. Soc. Jauary **14**, 14–22 (1913).

[BN] S. Bolotin and P. Negrini, *Variational approach to second species periodic solutions of Poincaré of the 3-body problem*, Discrete and Continuous Dynamical Systems, Volume 33, Number 3, March 2013, pp. 1009–1032.

[Boh] K. Bohlin, *Über eine neue Annäherungsmetode in der Störungsteorie*, Bihang till K. Svenska Vet.-Akad. **14** Afd. I, No. 5, Norstedt, Stockholm, 1888; partial republication in 1889 in Astronomische Nachrichten.

[Bo] J.L. Borges, Ficciones, Editorial Sur, Buenos Aires (1944).

[BH] B. Bramham and H. Hofer, *First steps towards a symplectic dynamics*, Surveys in Differential Geometry **17**, Number 1 (2012).

[Br] A.D. Bruno, *The restricted 3-Body Problem: Plane Periodic Orbits*, de Gruyter (1994).

[Bru] H. Bruns, *Über die Integrale des Vielkörper-problems*, Acta Mathematica **11** (1887).

[CMS] C. Carminati, S. Marmi and D. Sauzin, *There is only one KAM curve*, preprint 2011, `http://hal.archives-ouvertes.fr/hal-00585973/`.

[Ca] E. Cartan, *Leçons sur les invariants intégraux*, Hermann, Paris 1922.

[Cas] G. Casale, *Morales–Ramis theorems via Malgrange pseudogroup*, Ann. Inst. Fourier (Grenoble) **59**, n° 7, 2593–2610 (2009).

[CC1] A. Celletti and L. Chierchia, *On the stability of realistic Three-Body Problems*, Comm. Math. Phys. **186**, n° 2, 413–449 (1997).

[CC2] A. Celletti and L. Chierchia, *KAM stability and celestial mechanics*, Mem. Amer. Math. Soc. **187**, n° 878 (2007).

[Chap] M. Chaperon, *Une idée du type 'géodésiques brisées' pour les systèmes hamiltoniens*, C. R. Acad. Sci. Paris Sér. I Math. **298**, n° 13, 293–296 (1984).

[Cha] C.L. Charlier, *Die Mechanik des Himmels*, de Gruyter 1927.

[C0] A. Chenciner, *Systèmes dynamiques différentiables*, Encyclopédie Universalis.

[C1] A. Chenciner, *De la mécanique céleste à la théorie des systèmes dynamiques, aller et retour : Poincaré et la géométrisation de l'espace des phases*, Actes de la conférence "Epistémologie des systèmes dynamiques", Paris, décembre 1999, in *Chaos et systèmes dynamiques : Eléments pour une épistémologie*, S. Franceschelli, M. Paty & T. Roque eds., Hermann (2007).

[C2] A. Chenciner, *Une note de Poincaré*, Proceedings Curs Poincaré Barcelona, 2005, English translation *A note by Poincaré*, Regular and Chaotic Dynamics **10** n° 2, 119–128 (2005).

[C3] A. Chenciner, *Une promenade dans les Méthodes Nouvelles de la Mécanique Céleste*, Gazette des Mathématiciens n° 134, 37–47 (octobre 2012).

[C4] A. Chenciner, *Three-Body Problem*, Scholarpedia, `http://www.scholarpedia.org/article/Three_body_problem`.

[C5] A. Chenciner, *Four Lectures on the N-body problem*, in Hamiltonian Dynamical Systems and Applications, W. Craig ed., Springer (2008), 21–52.

[C6] A. Chenciner, *Compte-Rendu de lecture du premier volume des œuvres complètes de Jean Le Rond D'Alembert (premiers textes de mécanique céleste* 1747–1749), Gazette des Mathématiciens n° 99, 107–114 (janvier 2004).

[C7] A. Chenciner, *À l'infini en temps fini*, Sém. Bourbaki, exposé n° 832, juin 1997. `http://www.numdam.org/numdam-bin/feuilleter?id=SB_1996-1997__39_`.

[C8] A. Chenciner, *Solutions du problème des n corps joignant deux configurations : l'idée de Christian Marchal et ce qui s'en suit*, Gazette des Mathématiciens n° 99, 5–12 (janvier 2004).

[C9] A. Chenciner, *La dynamique au voisinage d'un point fixe elliptique conservatif: de Poincaré et Birkhoff à Aubry et Mather*, Sém. Bourbaki, 622, février 1994 `http://www.numdam.org/numdam-bin/feuilleter?id=SB_1983-1984__26_`.

[Chi] L. Chierchia, *The planetary N-body problem*, in Celestial Mechanics, edited by A. Celletti, Vol. 6.119.55 (2012) Encyclopedia of Life Support Systems (EOLSS), Developed under the Auspices of the UNESCO, Eolss Publishers.

[CP] L. Chierchia and G. Pinzari, *The planetary N-body problem: symplectic foliation, reductions and invariant tori*, Invent math **186**, 1–77 (2011) .

[Com1] T. Combot, *Non-intégrabilité algébrique et méromorphe de problèmes de n corps et de potentiels homogènes de degré -1*, Thèse Paris, décembre 2012.

[Com2] T. Combot, *A note on algebraic potentials and Morales–Ramis theory*, preprint 2012.

[Co] C. Conley, *On Some New Long Periodic Solutions of the Plane Restricted Three-body Problem*, Comm. Pure and Appl. Math. **XVI**, 449–467 (1963). See also [C5] (Section 1) and *Le problème de la lune et la théorie des systèmes dynamiques*, a course given by the author in 1985, available at `http://www.imcce.fr/Equipes/ASD/person/chenciner/chen_preprint.php`.

[CZ] C. Conley & E. Zehnder, *The Birkhoff–Lewis fixed point theorem and a conjecture of V.I. Arnold*, Invent. Math. **73**, 33–49 (1983).

[D] G. Darboux, *Éloge historique de Henri Poincaré*, lu dans la séance publique annuelle du 15 démbre 1913, Académie des Sciences, Institut de France.

[Da] G. Darwin, *Periodic orbits*, Acta Mathematica **21**, 99–242 (1897).

[En] *Encyclopédie ou dictionnaire raisonné des sciences, des arts et des métiers*, volume 9, article Lune (1765 but this article was written by d'Alembert in 1759).

[Fe1] J. Féjoz, *Global Secular Dynamics in the Planar Three-body Problem*, Celestial Mechanics and Dynamical Astronomy **84**, 159–195 (2002).

[Fe2] J. Féjoz, *The N-body problem*, in Celestial Mechanics, edited by A. Celletti, Vol. 6.119.55 (2012), Encyclopedia of Life Support Systems (EOLSS), Developed under the Auspices of the UNESCO, Eolss Publishers.

[Fe3] J. Féjoz, *On Arnold's theorem in celestial mechanics – a summary with an appendix on the Poincaré coordinates*, preprint (2012).

[Fe4] J. Féjoz, *Démonstration du "théorème d'Arnold" sur la stabilité du système planétaire (d'après Michael Herman)*, Michael Herman Memorial Issue, Ergodic Theory Dyn. Sys. **24**, 5, 1521–1582 (2004).

[FGKR] J. Féjoz, M. Guardia, V. Kaloshin, P. Roldan, *Diffusion along mean motion resonance in the restricted planar three-body problem*, preprint 2001, `http://www.ceremade.dauphine.fr/~fejoz/articles.php`.

[Fey] R.P. Feynman, *Lectures in Physics*, Vol. I, Addison Wesley 1963.

[GaK] J. Galante and V. Kaloshin, *Destruction of invariant curves in the restricted circular planar Three-Body Problem by using comparison of action*, Duke Math. J. **159**, Number 2, 275–327 (2011). Related preprints in `http://www2.math.umd.edu/~vkaloshi/`.

[GaU] J. Galante and J. Urschel, *Instabilities in the Sun-Jupiter-Asteroid Three Body Problem*, to appear in Celestial Mechanics and Dynamical Astronomy.

[GSV] V. Gelfreich, C. Simó and A. Vieiro, *Dynamics of 4D symplectic maps near a double resonance*, preprint `http://www.maia.ub.es/dsg/2012/` (number 5).

[Gh] E. Ghys, *Variations autour du théorème de récurrence de Poincaré*, Journal des élèves de l'ENS Lyon n° 1 (1994), available at `http://www.umpa.ens-lyon.fr/JME/Vol1Num1/artEGhys/artEGhys.html`.

[Go] E. Goursat, *Sur les transformations isogonales en Mécanique*, Comptes Rendus des Séances de l'Académie des Sciences, Paris, 108, 446–448 (1889).

[Gu] M. Gutzwiller, *Moon-Earth-Sun: The oldest Three-Body Problem*, Reviews of Modern Physics **70**, n° 2, (April 1998).

[Ha] J. Hadamard, *Les surfaces à courbures opposées et leurs lignes géodésiques*, J. Math. Pures Appl. **4**, 27–73 (1898).

[HK] Hasselblatt and A. Katok, Modern theory of Dynamical Systems, Cambridge University Press, 1996.

[He] M. Hénon, *Generating families in the Restricted Three-body Problem*, Lecture Notes in Physics: Monographs 52, Springer 1997.

[Her] M. Herman, *Some open problems in dynamical systems*, ICM Berlin 1998, II, 797–808.

[JT] E. Julliard-Tosel, *Bruns theorem: the proof and some generalizations*, Celestial Mechanics and Dynamical Astronomy **76**, 241–281 (2000).

[K] A.N. Kolmogorov, *On the Conservation of Conditionally Periodic Motions under Small Perturbations of the Hamiltonian*, Dokl. akad. nauk SSSR **98**, 527–530 (1954).

[Koz] V.V. Kozlov, *Integrability and non-integrability in Hamiltonian mechanics*, Uspechi Mat. Nauk 38:1 (1983), 3–67, English translation in Russian Math. Surveys **38**, 1, 1–76 (1983).

[Las1] J. Laskar, *Le système solaire est-il stable ?*, Séminaire Poincaré XIV (2010) 221–246. `http://www.bourbaphy.fr/laskar.pdf`. Is the Solar System Stable?, in Chaos, Poincaré Seminar 2010, B. Duplantier and V. Rivasseau eds., Progress in Mathematical Physics, Springer Basel AG 2013 (to appear).

[Las2] J. Laskar, *Poincaré et le Chaos dans le Système Solaire*, a conference given in Nancy, September 20 (2012). `http://www.poincare.fr/actualite/evenements?task=view_event&event_id=8`.

[LasRob] J. Laskar & P. Robutel, *Stability of the planetary Three-Body Problem, I. Expansion of the planetary Hamiltonian*, Cel. Mech. **62**, 193–217 (1995).

[Laz] V.F. Lazutkin, *The existence of caustics for a billiard problem in a convex domain*, Math. USSR Izv. **7**, 185–214 (1973).

[Le] P. Le Calvez, *Propriétés dynamiques des difféomorphismes de l'anneau et du tore*, Astérisque 204 (1991); English transl. *Dynamical properties of diffeomorphisms of the annulus and of the torus* SMF/AMS Texts & Monographs, 4.

[LC] T. Levi-Civita, *Sur la régularisation du problème des trois corps*, Acta Math. **42**, 99–144 (1920).

[LS] J. Llibre and C. Simó, *Oscillatory Solutions in the Planar Restricted Three-Body Problem*, Math. Ann. **248**, 153–184 (1980).

[Lya] A. Lyapunov, Obchtchaia zadatcha ob ouctoïtchivosti dvijenia, Kharkov 1892, French translation under the name of A. Liapounoff *Problème général de la stabilité du mouvement*, Annales de la faculté des sciences de Toulouse 2ème série, tome 9 (1907), p. 203–474, available at `http://afst.cedram.org/item?id=AFST_1907_2_9__203_0`, English translation *General Problem of the Stability of Motion*, CRC Press (1992).

[Mal] B. Malgrange, *On nonlinear differential Galois theory*, dedicated to the memory of Jacques-Louis Lions, Chinese Ann. Math. Ser. B **23**, no. 2, 219–226 (2002).

[MRL] F. Malige, P. Robutel, J. Laskar, *Partial reduction in the N-body planetary problem using the angular momentum*, Cel. Mech. Dyn. Astron. **84**, 283–316 (2002).

[Ma1] C. Marchal, *The Three-Body Problem*, Elsevier (1990).

[Ma2] C. Marchal, *How the method of minimization of action avoids singularities*, Celestial Mech. Dynam. Astronom. **83**, no. 1–4 (2002).

[McG] R. McGehee, *Singularities in Classical Celestial Mechanics*, Proceedings of the International Congress of Mathematicians, Helsinki 1978, 827–834.

[Moe1] R. Moeckel, *Some Qualitative Features of the Three-Body Problem*, Contemporary Mathematics **81**, 1–22 (1988).

[Moe2] R. Moeckel, *A Variational Proof of Existence of Transit Orbits in the Restricted Three-Body Problem*, Dynamical Systems: an International Journal **20**, 1, 45–58 (2005).

[Mont1] R. Montgomery *A New Solution to the Three-Body Problem*, Notices of the American Mathematical Society, 471–481, May. (2001).

[Mont2] R. Montgomery *N-body choreographies*, Scholarpedia, `http://www.scholarpedia.org/article/N-body_choreographies`.

[MS] J.J. Morales-Ruiz and S. Simon, *On the meromorphic non-integrability of some N-body problems*, DCDS-A **24** (4), 1225–1273 (2009).

[MR] J.J. Morales-Ruiz and J.P. Ramis, *Galoisian obstructions to integrability of Hamiltonian systems. I, II*, Methods Appl. Anal. **8** n° 1, 33–95, 97–111 (2001).

[MRS] J.J. Morales-Ruiz, J.P. Ramis, and C. Simó, *Integrability of Hamiltonian systems and differential Galois groups of higher variational equations*, Annales scientifiques de l'École normale supérieure **40** (6), 845–884 (2007).

[M1] J. Moser, *On invariant curves of area preserving mappings of an annulus*, Nachr. Akad. Wiss. Gött., Math. Phys. Kl. (1962), 1–20.

[M2] J. Moser, *Convergent Series Expansions for Quasi-Periodic Motions*, Math. Annalen **169**, 136–176 (1967).

[M3] J. Moser, *Stability theory in celestial mechanics*, in *The stability of the solar system and of small stellar systems*, p. 1–9, Kozai ed., I.A.U. (1974).

[M4] J. Moser, *Is the Solar System Stable?*, The Mathematical Intelligencer **1**, 65–71 (1978).

[M5] J. Moser, *Stable and random Motion in dynamical systems*, Princeton University Press (1973).

[Na] P. Nabonnand, *Les premières contributions d'Henri Poincaré en mécanique céleste vues à partir de sa correspondance avec Anders Lindstedt (1883- 1884)*, a lecture at the Bureau des Longitudes, Paris, October 3, 2012, HAL-00743645.

[Ne] N.N. Nekhoroshev, *An exponential estimate of the time of stability of nearly integrable Hamiltonian systems*, II. Trudy Sem. Petrovsk. **5**, 5–50 (1979).

[N] Sir I. Newton, *Opticks or a Treatise on the Reflections, Refractions, Inflections and Colours of Light.*

[Nie] L. Niedermann, *Stability over exponentially long times in the planetary problem*, Nonlinearity **9** (6), 1703–1751 (1996).

[Pa] P. Painlevé, *Leçons sur la théorie des équations différentielles professées à Stockholm* (*septembre, octobre, novembre* 1895), Hermann, Paris, 1897.

[P1] H. Poincaré, *Sur le problème des trois corps et les équations de la dynamique*, withdrawn version 1889 and Acta Mathematica **13**, 1–270 (1890).

[P2] H. Poincaré, *Les méthodes nouvelles de la mécanique céleste*, (1892, 1893, 1899), Gauthier-Villars, Paris; Russian translation with commentaries by Arnold, Alexeiev, Merman, Pogrebiskii, Moscow 1971; English translation with commentaries by D. Goroff, *New Methods of Celestial Mechanics*, American Institute of Physics (1992).

[P3] H. Poincaré, *Mémoire sur les courbes définies par une équation différentielle*, 4 parts 1881, 1882, 1885, 1886, in Œuvres, tome I (the title of the two last parts is *Sur les courbes définies par les équations différentielles*).

[P4] H. Poincaré, *Sur certaines solutions particulières du problème des trois corps*, Bulletin astronomique **1**, 65–74 (février 1884).

[P5] H. Poincaré, *Sur les solutions périodiques et le principe de moindre action*, C.R.A.S. **123**, 915–918 (1896); in Œuvres, tome VII.

[P6] H. Poincaré, *Le problème des trois corps*, Revue générale des sciences pures et appliquées **2**, 1–5 (1891).

[P7] H. Poincaré, *Sur les petits diviseurs dans la théorie de la Lune*, Bulletin astronomique. **25**, 321–360 (1908), [Œuvres tome 8. p. 332–366].

[P8] H. Poincaré, *Sur la stabilité du système solaire*, *Annuaire pour l'an 1898 du Bureau des Longitudes*, Gauthier-Villars et fils 1897, reproduit dans la Revue scientifique (revue rose), 4'eme série, tome IX, 20, 14 mai 1898.

[P9] H. Poincaré, *Sur la méthode de Bruns*, C.R.A.S. **123**, 1224–1228 (1896).

[P10] H. Poincaré, *Les solutions périodiques et le principe de moindre action*, C.R.A.S. **124**, 713–716 (1897).

[P11] H. Poincaré, *Sur la façon de grouper les termes des séries trigonométriques qu'on rencontre en Mécanique céleste*, Bulletin astronomique **15**, 289–310, (août 1898).

[P12] H. Poincaré, *Leçons de mécanique céleste*, Gauthier-Villars (1905, 1907, 1910).

[P13] H. Poincaré, *La valeur de la science*, Bibliothèque de philosophie scientifique, Flammarion 1905; English translation *The value of science*, Dover 1958.

[P14] H. Poincaré, Sur les lignes géodésiques des surfaces convexes, Transactions of the American Mathematical Society **6**, 237–274 (july 1905).

[P15] H. Poincaré, *Sur un théorème de géométrie*, Rendiconti del Circolo matematico di Palermo **33**, 375–407 (1912).

[PC] *La Correspondance de Henri Poincaré, Vol. 3 : La Correspondance entre Henri Poincaré et les astronomes et géodésiens*, Ralf Krömer, Philippe Nabonnand, Martina Schiavon & Scott Walter eds., Basel: Birkhäuser, to appear in 2013.

[Pos] J. Pöschel, *Integrability of Hamiltonians on Cantor sets*. Communications in Pure and Applied Math. **XXXV**, 653–695 (1982).

[Ra1] J.P. Ramis, *Poincaré et les développements asymptotiques*, (1ère*partie*), Gazette des mathématiciens **133**, 33–72 (juillet 2012).

[Ra2] J.P. Ramis, *Les développements asymptotiques après Poincaré : continuité et ... divergences*, Gazette des mathématiciens **134**, 17–36 (octobre 2012).

[Ro] A. Robadey, *Différentes modalités de travail sur le général dans les recherches de Poincaré sur les sytèmes dynamiques*, Thèse Paris 2006.

[Rob] P. Robutel, *Stability of the planetary Three-Body Problem II, KAM theory and existence of quasi-periodic motions*, Celestial Mech. Dynam. Astron. **62**, 219–261 (1995).

[Sau] D. Sauzin, *Résurgence paramétrique et exponentielle petitesse de l'écart des séparatrices du pendule rapidement forcé*, Thèse, Paris 1994.

[SM] C.L. Siegel and J.K. Moser, *Lectures on Celestial Mechanics*, Grundlehren der mathematischen Wissenschaften. Springer Verlag (1971).

[Si] C. Simó, *Some properties of the global behaviour of conservative low dimensional systems*, in *Foundations of Computational Mathematics: Hong Kong* 2008, F. Cucker et al. editors, pp. 163–189, London Math. Soc. Lecture Notes Series **363**, Cambridge Univ. Press, 2009.

[SST] C. Simó, P. Sousa-Silva and M. Terra, *Domains of Practical Stability near* $L4, 5$ *in the 3D Restricted Three-Body Problem*, Slides of the lecture given at Dynamical Systems: 100 years after Poincaré, Gijón (Spain) 3–7 September 2012, `http://www.maia.ub.es/dsg/2012/` (number 8).

[ST] C. Simó and T.J. Stuchi, *Central stable/unstable manifolds and the destruction of KAM tori in the planar Hill problem*, Physica D: Nonlinear Phenomena **140**, Issues 1–2, 1–32 (1 June 2000).

[Sin] Ya.G. Sinai, *Introduction to ergodic theory*, Erevan state University (in Russian) 1973, English transl.: Mathematical Notes, Princeton Univ. Press, 1977.

[Sm1] S. Smale, *Diffeomorphisms with many periodic points*, Differential and Combinatorial Topology, Princeton Univ. Press, 1965, pp. 63–80.

[Sm2] S. Smale, *Differentiable Dynamical Systems*, Bull. Amer. Math. Soc. **73**, 747–817 (1967).

[Sz] V. Szebehely, *The Theory of Orbits*, Academic Press (1967).

[Th] R. Thom, *Prédire n'est pas expliquer*, Eshel, Paris (1991).

[Ti] Tisserand, *Traité de Mécanique céleste*, Gauthier-Villars, vol 1: 1889, vol. 2: 1891, vol. 3: 1894, vol. 4: 1896.

[V] C. Villani, *(Ir)réversibilité et entropie*, Séminaire Poincaré XV, Le temps (2010). `http://www.bourbaphy.fr/villani.pdf`.
(Ir)reversibility and Entropy, in Time, Poincaré Seminar 2010, B. Duplantier ed., Progress in Mathematical Physics, vol. 63, pp. 19–79, Springer Basel AG (2013). *(Ir)réversilbilité et entropie*, in Time, Poincaré Seminar 2010, B. Duplantier ed., Progress in Mathematical Physics, vol. 63, pp. 81–144, Springer Basel AG (2013).

[Wi] A. Wintner, *The Analytical Foundations of Celestial Mechanics*, Princeton University Press (1949).

[Y] J.C. Yoccoz, *Une erreur féconde du mathématicien Henri Poincaré*, Gazette des mathématiciens, n° 107, p. 19–26 (2006).

[Ze] E. Zehnder, *Homoclinic Points Near Elliptic Fixed Points*, Communications on pure and applied mathematics **XXVI**, 131–182 (1973).

[Z] S.L. Zyglin, *Branching of solutions and nonexistence of first integrals in Hamiltonian mechanics, i*, Functional Analysis and Its Applications **16** (3), 181–189 (1982).

Alain Chenciner
Observatoire de Paris
IMCCE (UMR 8028), ASD
77, avenue Denfert-Rochereau
F-75014 Paris, France

and

Département de mathématique
Université Paris VII
e-mail: `chenciner@imcce.fr`

Henri Poincaré, 1912–2012, 151–192

Poincaré Seminar 2012

Poincaré's Odds

Laurent Mazliak

Abstract. This paper is devoted to Poincaré's work in probability. Though the subject does not represent a large part of the mathematician's achievements, it provides significant insight into the evolution of Poincaré's thought on several important matters such as the changes in physics implied by statistical mechanics and molecular theories. After having drawn the general historical context of this evolution, I focus on several important steps in Poincaré's texts dealing with probability theory, and eventually consider how his legacy was developed by the next generation.

Introduction

In 1906, Poincaré signed one of the most unusual texts of his scientific career [1], a report written for the Cour de Cassation in order to eventually close the Dreyfus case. In 1904, ten years after the arrest and the degradation of the unfortunate captain, the French government decided to bring an end to this lamentable story which had torn French society apart for years, and obtained the rehabilitation of the young officer who had been so unjustly martyred. As is well known, Dreyfus was convicted of treason in 1894 in the absence of any material proof, during a rushed and unbalanced trial orchestrated by the military hierarchy, whose aim was to identify a culprit as soon as possible. The only concrete document submitted as evidence by the prosecution was the famous *bordereau*, found in a wastepaper basket in the German Embassy in Paris and briefly scrutinized by several more or less competent experts. Among them, Alphonse Bertillon played a specially sinister role and became Dreyfus' most obstinate accuser. He built a bizarre, scientific-sounding theory of self-forgery of the *bordereau*, purporting to prove Dreyfus' guilt. Bertillon became trapped by his own theoretical construction, more because of overweening self-confidence and stupidity than by a real partisan spirit. When the Affair broke out at the end of the 1890s, and the political plot became obvious and Dreyfus' innocence apparent, Bertillon continued to elaborate his theory, rendering it more and more complex. This frenzy of elaboration brought him an avalanche of trouble, and nearly put an end to his career in the Paris police department. However, for the trial in the *Cour de Cassation*, in order to silence the last dissenting voices which might be raised, it was decided not to ignore Bertillon's rantings, but to ask incontestable academic authorities to give their opinion about the possible value of the self-proclaimed expert's conclusions. Three mathematicians were called for that purpose, Paul Appell, Gaston Darboux and Henri Poincaré, who jointly signed the report for the Cour de Cassation in 1906. Nevertheless, it was well known that only the latter worked on the document, a tedious task Poincaré that undertook with integrity, but somewhat grudgingly. Besides, this was not Poincaré's first intervention in the Dreyfus case: during the judicial review of 1899 in Rennes of the decision of 1894, Painlevé read a letter of Poincaré criticizing Bertillon's work for its lack of scientific foundation.

This story is well known and has been narrated in detail many times ([52], [53], [81]) and I shall not dwell on it further. However, one may ask why it was Poincaré who was called upon to carry out this task. The first answer that comes to mind is the following: in 1906, Poincaré, who was then fifty-two years old, was without a doubt the most prominent among French scientists. He had, moreover, another characteristic: his name was familiar to a sizeable audience outside the scientific community. It was well known that he had been awarded the important prize of the Swedish king Oscar II in 1889; the publication of popular books on the philosophy of science increased his name recognition, and he was also famous due to several essays published in newspapers, for instance during the International Congress of Mathematicians in Paris in 1900, over which Poincaré presided. To seek out such a person in order to crush the insignificant but noisy Bertillon was therefore a logical calculation on the government's part. However, a second and more hidden reason probably played a part as well. The report for the *Cour de Cassation* [1] opens with a chapter whose title was quite original in the judicial literature: *Notions on the probability of causes*; it contains a brief exposition of the principles of the Bayesian method. Bertillon had indeed pretended to build his system on the methods and results of the theory of probability, and answering him was only possible by confronting the so-called expert with his own weapons. It was therefore necessary not only to seek out a scientific celebrity for the job, but also someone whose authority in these matters could not be challenged. In 1906, Poincaré was regarded by everyone as the leading specialist in the mathematics of randomness in France. While by then he was no longer the Sorbonne Chair of Calculus of Probability and Mathematical Physics, he had held the Chair for some ten years (it was in fact the first position he obtained in Paris), had accomplished an impressive amount of work in mathematical physics during this period (we shall return at length to this subject later), and had published in 1896, shortly before leaving the chair, a treatise on the Calculus of Probability which, in 1906, was still the main textbook on the subject in French. Moreover, several texts had presented his thoughts on the presence of randomness in modern physics to a large cultivated audience, in particular his *Science and Hypothesis* ([71]), which went through several editions. It was therefore as a specialist in the calculus of probability that Poincaré was sought out by the judicial authorities, in order to dispose of the Dreyfus case once and for all.

If we go back in time fifteen years before this event, we notice a striking contrast. Since 1886, although Poincaré held the aforementioned Sorbonne chair, he had probably mainly seen in the chair's title the words *Mathematical Physics.* In 1892, he published an important textbook on Thermodynamics [64] based on the lectures he had read at the Sorbonne several years before. A Poincaré publication would of course not go unnoticed, and one attentive reader had been the British physicist Peter Guthrie Tait (1831–1901). Tait had been very close to Maxwell and was one of the most enthusiastic followers of his work. He wrote a review of Poincaré's book for the journal *Nature* ([87]); the review was quite negative despite the obvious talent Tait recognized in his young French colleague.

Poincaré, wrote Tait, introduced beautiful and complex mathematical theories in his textbook but often to the detriment of the physical meaning of the situations he studied. The most important reproach of the British physicist was the fact that Poincaré remained absolutely silent about the statistical theories of thermodynamics, leaving in the shadows the works of Tait's friend and master Maxwell. Tait wrote:

> 'But the most unsatisfactory part of the whole work is, it seems to us, the entire ignoration of the true (i.e., the statistical) basis of the Second Law of Thermodynamics. According to Clerk–Maxwell (Nature, xvii. 278) "The touch-stone of a treatise on Thermodynamics is what is called the Second Law." We need not quote the very clear statement which follows this, as it is probably accessible to all our readers. It certainly has not much resemblance to what will be found on the point in M. Poincaré's work: so little, indeed, that if we were to judge by these two writings alone it would appear that, with the exception of the portion treated in the recent investigations of v. Helmholtz, the science had been retrograding, certainly not advancing, for the last twenty years.'

Poincaré wrote an answer and sent it to *Nature* on 24 February 1892. It was followed, during the first semester of 1892, by six other letters between Tait and Poincaré[1]; they are rather sharp, each sticking to his position. On March 17th, Poincaré wrote the following comment about the major criticism made by Tait:

> 'I left completely aside a mechanical explanation of the principle of Clausius that M. Tait calls "the true (i.e., statistical) basis of the Second Law of Thermodynamics." I did not speak about this explanation, which besides seems to me rather unsatisfactory, because I desired to remain absolutely outside all the molecular hypotheses, as ingenious they may be; and in particular I said nothing about the kinetic theory of gases.'[2]

[1] Accessible freely on the website of the Archives Poincaré in Nancy at the address http://www.univ-nancy2.fr/poincare/chp/hpcoalpha.xml

[2] 'J'ai laissé complètement de côté une explication mécanique du principe de Clausius que M. Tait appelle "the true (i.e. statistical) basis of the Second Law of Thermodynamics." Je n'ai pas parlé de cette explication, qui me paraît d'ailleurs assez peu satisfaisante, parce que je désirais rester complètement en dehors de toutes les hypothèses moléculaires quelque ingénieuses qu'elles puissent être; et en particulier j'ai passé sous silence la théorie cinétique des gaz'.

One thus observes that Poincaré, in 1892, had a very negative view of statistical mechanics, the very science of matter in which probabilities played the greatest role. However, Poincaré would not have been Poincaré if, once shown a difficulty, he did not take the bull by the horns and to try to tame it. A first decision was the resolution, during the next academic year, 1893, to lecture on the kinetic theory of gases. And indeed, in the Sorbonne syllabus for that year, we see that Poincaré had transformed his lectures into *Thermodynamics and the kinetic theory of gases.*

In 1894 Poincaré's first published paper on the kinetic theory of gases [67], to which we shall return later. If one still observes a good deal of skepticism in it, or at least reservation about these novelties in randomness, a change of attitude was taking place. This same academic year 1893–1894, Poincaré eventually decided for the first time to teach a course on the calculus of probability at the Sorbonne, which subsequently became the basis of his book of 1896, the same year when he exchanged his chair for that of celestial mechanics. Moreover, in the following years, reflections on the mathematics of randomness became more frequent in his writings, up to the publication of his more philosophical works, which acknowledged the integration of the theory of probability among Poincaré's mathematical tools. By 1906, as already noted, the transition had been completed, especially as new elements, such as Einstein's just published theory of Brownian motion, made even more necessary the increasing presence of probability in scientific theories.

The present paper concerns the probabilistic aspects of Poincaré's enormous production, aspects of which remain somewhat limited in size. A non-negligible challenge in dealing with this subject is that probability penetrated Poincaré's work almost by force, forcing his hand several times, his main achievement consisting in building dykes so that the mathematician might venture with a dry foot on these rather soft lands. We shall see later that his successor Borel had a somewhat different attitude toward choosing to apply probability theory in many domains, reflecting the fact that Borel had encountered probability in a more spectacular way than Poincaré.

Despite this limited contribution, Poincaré succeeded in leaving a significant heritage which would later prove important. Above all, his most decisive influence may have been to allow probability theory to regain its prestige in France again, after occupying a rather miserable position in the French academic world for more than half a century.

The aim of the present paper is therefore threefold, and this is reflected in its three parts. In the first part, we focus on Poincaré's evolution in the fifteen years we have just pointed to, in order to define how this progressive taming of the mathematics of randomness occurred. In the second section, I examine in detail several of Poincaré's works involving probability considerations in order to give an idea not only of Poincaré's style, but also of the intellectual basis for the evolution in his thinking. And finally, in the last section, I comment on the persons who recovered the heritage of Poincaré in this area, and on how they extended it.

Many of the topics discussed in this paper have been covered previously. In particular, I shall refer several times to Sheynin's text ([84]), to von Plato's publications ([90], [91]) as well as to several papers by Bru, some published and some not ([20], [23]). Let us me also mention the nice book [36] where the author deals with Poincaré's works on probability with an epistemological approach.

Acknowledgement

I warmly thank the organizers of the Séminaire Poincaré on 24 November 2012 in Paris, for the opportunity to end the Poincaré year 2012 with a closer reading of his work on probability, work which until then I knew only superficially. I hope to have succeeded in the paper to present my understanding of a great mind's intellectual peregrination, which led him to such a considerable evolution in his thought on the subject. Special thanks go to Bertrand Duplantier who made many comments on a first version of the paper. Next, I thank Xavier Prudent who helped me to read the 1894 paper on the kinetic theory of gases; needless to say, the remaining obscurities on the topic are entirely mine. It is also a pleasure to express all my gratitude to Sandy Zabell and Scott Walter for their generous help in editing the English of the present paper; here also I take the complete responsibility for all the infelicities which may remain in the text. After the presentation of this paper at the Institut Henri Poincaré on 24 November 2012, I had an interesting echange with Persi Diaconis; I hope he will find the paper improved by his comments. And finally, I feel it necessary to say how much Bernard Bru would have deserved to write this chapter. I dedicate this work with gratitude to the memory of an honest man, our master and friend Marc Barbut, whom we miss so much since a cold morning of December 2011...

1. First part: the discovery of probability

In the beginning was the Chair. For, as we will see, it was not only by a fortuitous scientific interest that Poincaré first came to be involved in probability theory, but there was also a very specific academic situation in which Hermite, the major mathematical authority in 1880s France, crusaded for the career of his three *mathematical stars*: Paul Appell (his nephew), Émile Picard (his son-in-law) and of course Henri Poincaré, the latter considered by him to be the most brilliant, though he did not belong to his family (to the great displeasure, so he wrote to Mittag-Leffler, of Madame Hermite). A remarkable change took place at the Sorbonne at this time when, in just a few years, almost every holder of the existing chairs in mathematics and physics died: Liouville and Briot in 1882, Puiseux in 1883, Bouquet and Desains in 1885, Jamin in 1886, leaving an open field for a spectacular change in the professorships. At the end of the whirlwind, the average age of the professors of mathematics and physics at the Sorbonne had been reduced by eighteen years! It was therefore clear for everyone, and first of all Hermite, that there was a risk that the situation might become fixed for a long time, and that it was therefore necessary to act swiftly and resolutely in favor of his protégés. The three of them were, indeed, appointed in Paris during these years. This episode is narrated in detail in [2] and I here touch only upon the most important aspects insofar as they relate to our story.

The Chair of Calculus of Probability and Mathematical Physics, occupied until 1882 by Briot, had been created some thirty years before, after numerous unsuccessful attempts by Poisson, but in a form different than the one pursued by the latter. The joining of mathematical physics to probability had been decided in order to temper the bad reputation of the theory of probability in the 1840s in France, notably due to some of the work of Laplace and, above all, of Poisson himself dealing with the application of probability in the judicial domain. The book of Poisson [79] ignited a dispute in the Academy in 1836, when Dupin and Poinsot harshly contested Poisson's conclusions, and the philosophers led by Victor Cousin made a scene *in the name of the sacred rights of liberty against the claims advanced by the mathematicians in order to explain how events occur in social issues.* John Stuart Mill summed up the situation by qualifying the application of probability to judicial problems as the *scandal of mathematics.* This polemical contretemps tarred the calculus of probability in France, leaving it with a most dubious reputation.

Although Lippmann was nominated in 1885 for the chair, Jamin's unexpected demise freed the chair of Research in physics of which Lippmann took hold immediately, leaving a place for Poincaré. One might initially be surprised by this choice. First, because Poincaré was preferred to other candidates, of course less gifted, but whose area of research was closer to that of the position. One would rather have expected to see Poincaré in one of the chairs in Analysis, for instance. In 1882, when Briot died and the great game began, Poincaré had no published work in physics (let alone the calculus of probability). And yet, among the candidates who were eliminated, there was for instance Boussinesq who had to his credit significant contributions in that domain, and who openly stated his intention of revitalizing the teaching of probability which seemed to him to be in a poor state. Not without irony, Boussinesq later became Poincaré's successor in this same chair when the latter changed over to the Chair of Celestial Mechanics.

Thus Poincaré was nominated in 1886, without any real title for the position, and one may think that it had really been chance which led him there. However, as shown in Atten's fine analysis [2], a number of indications demonstrate that Poincaré had indeed really desired this particular position. Examining his lectures in 1887–1888 shows that he already had a profound knowledge of physical theories and, moreover, several passages in his correspondence at this time show his sustained interest in questions of physics. This shows that his attitude had not been purely opportunistic and that the chair pleased him. Also, Hermite, apart from the desire of supporting his protégé, seems to have made a thoughtful bet in nominating him for this somewhat unexpected position. Knowing Poincaré's acute mind, it was not unreasonable to anticipate spectacular achievements by him in this position. What followed, as we know, bore out Hermite's expectation...

The theory of probability, as I said, did not seem to concern the new professor at the beginning of his tenure, and he taught courses on several different physical theories. In 1892, he published his lectures on thermodynamics, delivered in 1888–1889 [64], producing a book, as noted earlier, that was sharply criticized by Tait. Poincaré therefore decided to look into the questions raised by the kinetic theory of gases, in particular because he had just read a communication by Lord Kelvin to the Royal Society containing several fundamental criticisms on Maxwell's theory [47]. Perhaps Poincaré had been especially eager to read this paper because it might provide a powerful argument in his controversy with Tait. However, the affair took another direction, revealing the mathematician's profound scientific honesty. At the beginning of the paper [67], published in 1894, though he again expressed some skepticism, Poincaré, whose conventionalism was formed during these years, seemed already half convinced of the possible fecundity of Maxwell's theory.

> 'Does this theory deserve the efforts the English devoted to it? One may sometimes ask the question; I doubt that, right now, it may explain all the facts we know. But the question is not to know if it is true; this word does not have any meaning when this kind of theory is concerned. The question is to know whether its fecundity is exhausted or if it can still help to make discoveries. And admittedly we cannot forget that it was useful to M. Crookes in his research on radiant matter and also to the inventors of the osmotic pressure. One can therefore still make use of the kinetic hypothesis, as long as one is not fooled by it.'[3]

We shall come back in the second section to this article of 1894 and to the new formulation of the ergodic principle it contains, in which Poincaré introduced the restriction of exceptional initial states. Let us only mention here the following point: Poincaré seemed to have found this idea in his previous work on the three body problem for which he had obtained the prize of the King of Sweden in 1889. In the memoir presented for the prize, he had indeed proved a recurrence theorem concerning the existence of trajectories such that the system comes back an infinite number of times in any region of the space, no matter how small. The next year, for his paper in *Acta Mathematica*, Poincaré added a probabilistic extension of his theorem where he showed that the set of initial conditions for which the trajectories come back only a finite number of times in the selected region has probability zero ([62], pp. 71–72); this passage was included some years later in his treatise of new methods for celestial mechanics [65] in a section simply called 'Probabilities'.

During the academic year 1893–1894, Poincaré prepared his first course of probability for his students of mathematical physics ([69]), published in 1896 and transcribed by Albert Quiquet, a former student of the École Normale Supérieure

---

[3]'Cette théorie mérite-t-elle les efforts que les Anglais y ont consacrés ? On peut quelquefois se le demander; je doute que, dès à présent, elle puisse rendre compte de tous les faits connus. Mais il ne s'agit pas de savoir si elle est vraie; ce mot en ce qui concerne une théorie de ce genre n'a aucun sens. Il s'agit de savoir si sa fécondité est épuisée ou si elle peut encore aider à faire des découvertes. Or, on ne saurait oublier qu'elle a été utile à M. Crookes dans ses travaux sur la matière radiante ainsi qu'aux inventeurs de la pression osmotique. On peut donc encore se servir de l'hypothèse cinétique, pourvu qu'on n'en soit pas dupe.' ([67], p. 513)

who entered the institution in 1883, before becoming an actuary, and who had probably been an attentive listener of the master's voice. In the manner of the standard probability textbooks of the time, it does not present a unified theoretical body but rather a series of questions that Poincaré tried to answer (the main ones, which occupy the bulk of the volume, concern the theory of errors of measurement - a subject we shall return to in the next section). The book, in its edition of 1896, is a natural successor to Bertrand's textbook [9], which up to then was the standard textbook (see [21]). In comparison to Bertrand's book, Poincaré's consolidates the material in several interesting ways, and this aspect is even more obvious in the second edition [77], completed by Poincaré some months before his death in July, 1912.

The mathematician seemed now convinced that it was no longer possible to get rid of probability altogether in science, so he decided instead to make the theory as acceptable as possible to the scientist. Poincaré decided to devote considerable effort towards that end, especially by writing several texts lying half-way between popularization (with the meaning of writing a description of several modern concepts using as little technical jargon as possible) and innovation. Two texts are of particular importance: that of 1899 ([70]) – reprinted as a chapter of [71] – and that of 1907 ([75]) – reprinted again as [76] and then as a preface to the second edition of his textbook [77] – two texts marking Poincaré's desire to show off his new probabilistic credo. But one has to realize that Poincaré wanted to convince himself above all and this leads one to ask the question Jean-Paul Pier ironically used as the title for his paper [61]: did Poincaré believe in the calculus of probability or not? Without pretending to give a final answer, one may however observe that Poincaré very honestly sought for a demarcation of the zone where it seemed to him that using probability theory did not create a major problem. Hence the attempt to tackle some fundamental questions in order to go beyond the defects that Bertrand had ironically illustrated with his famous paradoxes: When is it legitimate to let randomness intervene? What definition can be given of probability? Which mathematical techniques can be developed in order to obtain useful tools for physics, in particular for the kinetic theory of gases? Borel, as we shall see in the third section, would later remember Poincaré's position.

In his 1907 text, Poincaré accurately defined the legitimate way to call upon the notion of randomness. He saw essentially three origins for randomness: the ignorance of a very small cause that we cannot know but which produces a very important effect (such as the so-called Butterfly Effect), the complexity of the causes which prevents us from giving any explanation other than a statistical one (as in the kinetic theory of gases), the intervention of an unexpected cause that we have neglected. This was not too far from the Laplacian conception, which should not surprise us very much since Poincaré, born in 1854, was a child of a century for which Laplace was a tutelary figure. However, Poincaré knew well the accusations accumulated against Laplace's theory and he proposed several ways to adjust it: randomness, even if it is connected to our ignorance to a certain extent, is not only that, and it is important to define the nature of the connection between randomness and ignorance. The conventionalist posture on which we have already commented naturally made things easier, but Poincaré did not seek facility. As he wrote in 1899:

> 'How shall we know that two possible cases are equally probable? Will it be by virtue of a convention? If we state an explicit convention at the beginning of each problem, everything will be fine; we will just apply the rules of arithmetic and algebra, and pursue the computation to the end, such that our result leaves no room for doubt. But, if we want to apply it anywhere, we will have to prove the legitimacy of our convention, and we will find ourselves in front of the difficulty we thought we had eluded.'[4]

In a remarkable creative achievement, Poincaré forged a method allowing the objectification of some probabilities. Using it, if one considers for instance a casino roulette with alternate black and red sectors, even without having the slightest idea of how it is put into motion, one may show it is reasonable to suppose that after a large number of turns, the probability that the ball stops in a red zone (or a black zone) is equal to 1/2. There are thus situations where one can go beyond the hazy Laplacian principle of (in)sufficient reason as a necessary convention to fix the value of the probability. The profound *method of arbitrary functions*, which is based on the hypothesis that at the initial time the distribution of the position where the ball stops is arbitrary and shows that this distribution reaches an asymptotic equilibrium and tends towards the uniform distribution, was certainly Poincaré's most important invention in the domain of probability and we shall see later the spectacular course it took.

To conclude this survey, let us mention that in 1906, when Poincaré was completing his report for the Cour de Cassation in the Dreyfus case, despite his place as the pre-eminent French authority in the theory of probability, he was, as far as his scientific thought was concerned, somewhat in the middle of the ford between a completely deterministic description of the world and our modern conceptions where randomness enters as a fundamental ingredient. Poincaré kept this uncomfortable position until the end of his life. It is besides noticeable that at the precise moment when Borel was – so strikingly – taking over, Poincaré did not seem to have been particularly interested in the enterprise of his young follower. He showed similarly disinterest for the fortunate experiments of the unfortunate Bachelier: he had written, it is true, a benevolent report on his thesis [3] and had sometimes helped him to obtain grants, but the two men had no further scientific contact [29]. Even more surprising, Poincaré seems to have thoroughly ignored the Russian school's works (Chebyshev, Markov, Lyapounov...) and this

[4]'Comment saurons nous que deux cas possibles sont également probables? Sera-ce par une convention? Si nous plaçons au début de chaque problème une convention explicite, tout ira bien; nous n'aurons plus qu'à appliquer les règles de l'arithmétique et de l'algèbre et nous irons jusqu'au bout du calcul sans que notre résultat puisse laisser place au doute. Mais, si nous voulons en faire la moindre application, il faudra démontrer que notre convention était légitime, et nous nous retrouverons en face de la difficulté que nous avions cru éluder.' ([70], p. 262)

explains why he became never conscious of some connections with his own works. Poincaré's probabilistic studies leave therefore a feeling of incompleteness, partly due probably to his premature death at the age of 58, but also to the singular situation of this last giant of Newtonian-Laplacian science who remained on the threshold of upheavals that arrived after his death.

2. Second part: construction of a probabilistic approach

In this second part, I would like to present some steps which have marked the progressive entry of questions on probability in Poincaré's works. Even if it is not possible to see a perfect continuity in the chain of his research, a kind of genealogy can be traced which allows one to better understand how the mathematician gradually adopted a probabilistic point of view in several situations. I feel a need to mention that my aim in this section is to describe Poincaré's contribution and not necessarily to comment its originality. A particular problem with our hero is that he practically never quotes his sources so that it is difficult to make a statement about what he had read or not. I have tried as much as was possible to make each subsection of this part independent of the others, which may sometimes result in brief repetition.

2.1. The recurrence theorem and its 'probabilistic' extension

In anticipation of the sixtieth birthday of King Oscar II of Sweden in 1889, a mathematical competition was organized by Mittag-Leffler. The subject concerned the three-body problem: Was the system Earth-Moon-Sun stable? Periodic? Organized so that it will always remain in a finite zone of space? Many of these questions were fundamental ones that had challenged Newtonian mechanics from the 18th century. Poincaré submitted in 1888 an impressive memoir, immediately selected for the award by a jury including Weierstrass, Mittag-Leffler and Hermite. While correcting the proofs of the paper for *Acta Mathematica*, Phragmen located a mistake, leading Poincaré to make numerous amendments before resubmitting a lengthy paper the following year, published in volume 13 of *Acta Mathematica*.

This story, well known and well documented (see in particular [7]), is of interest for us only as far as one difference between the version submitted for the prize and the published version in 1890 is the appearance of the word *probability*, certainly for the first time in the French mathematician's works. I shall closely follow Bru's investigation [23] on the way in which countable operations gradually established themselves in the mathematics of randomness.

In the first part of the memoir submitted for the competition, Poincaré studied the implications of the existence of integral invariants on the behavior of dynamical systems. A simple example of this situation is given by the case of an incompressible flow for which the shape of the set of molecules changes, but not its volume, which remains constant in time. However, Poincaré had in view a much more general situation. He showed in fact on p. 46 *and seq.* of [62] that with a proper choice of coordinates in the phase space of position and velocity, the general situation of a mechanical system may be expressed by such an integral invariant. Poincaré then expounded a version of his recurrence theorem in the following form: let E be a bounded portion of the space, composed of mobile points following the equations of mechanics, so that the total volume remains invariant in time. Let us suppose, writes Poincaré, that the mobile points remain always in E. Then, if one considers r_0 a region of E, no matter however small it may be, there will be some trajectories which will enter it an infinite number of times.

Poincaré's proof is a model of ingenuity and simplicity. One discretizes time with a step of amplitude τ. Let us call with Poincaré

$$r_1, r_2, \dots, r_n, \dots$$

the "consequents" of r_0, that is to say, the successive positions of the different points of the region r_0 at times

$$\tau, 2\tau, \dots, n\tau, \dots$$

In the same way, the "antecedent" of a region is the region of which it is the immediate consequent. Each region r_i has the same volume; as they remain by hypothesis inside a bounded zone, some of them necessarily intersect. Let two such regions be r_p and r_q with $p < q$, with intersection the region s_1 having nonzero volume: a point starting in s_1 will be back in s_1 at time $(q-p)\tau$. Going back in time, let us call r_0^1 the sub-region of r_0 whose pth consequent is s_1. A particle starting from r_0^1 will again enter this region at time $(q-p)\tau$. We now start again the process by replacing r_0 by r_0^1, and thus build a decreasing sequence (r_0^n) of sub-regions of r_0 such that each point starting from r_0^n returns n times at least. Considering a point in the intersection of the r_0^n (whose non-emptiness is taken for granted by Poincaré), a trajectory starting from such a point will pass an infinite number of times through r_0.

In this form, the theorem is therefore completely deterministic. What then compelled Poincaré to believe it necessary, in the new version published in 1890, to rewrite his result in a probabilistic setting ([62], pp. 71–72)? At first glance indeed, the appearance of the word probability in these pages may seem surprising in the chosen framework of the mechanics of Newton, Laplace and Hamilton. In fact, as Bru remarks ([23]), one must not be misled by this use of probability by Poincaré, viewing it as a sudden revelation of the presence of randomness having the ontic value we spontaneously give it today. Poincaré himself wrote: *je me propose maintenant d'expliquer pourquoi [les trajectoires non récurrentes] peuvent être regardées comme exceptionnelles.* What Poincaré was thus looking for was a convenient way of expressing the rarity, the thinness of a set. He was writing before the decisive creation of the measure theoretic tools and particularly of Borel's thesis which would, four years later, prove that a countable set has a measure equal to zero. For a long while, astronomers in particular employed the concept of probability with the meaning of practical rarity and certainly one should not seek for a more sophisticated explanation to justify the presence of the word coming from Poincaré's pen. It was a convenient way of speaking, whose aim was almost entirely to hide the *obscure instinct* mentioned by Poincaré in his 1899 paper ([70], p. 262), almost as an apology, because we cannot do without it if we want to do scientific work.

Poincaré begins by expounding the following 'definition': if one calls p_0 the probability that the considered mobile point starts from a region r_0 with volume v_0 and p'_0 the probability that it starts from another region r'_0 with volume v'_0, then

$$\frac{p_0}{p'_0} = \frac{v_0}{v'_0}.$$

In particular, if r_0 is a region with volume v, used as a reference, the probability that the mobile point starting from r_0 starts from a sub-region σ_0 with volume w is given by $\frac{w}{rv}$. Equipped with this notion, the mathematician wants to prove that the initial conditions in r_0 such that the trajectory does not reenter r_0 more than k times form a set with probability zero, no matter how large the integer k.

Earlier in his paper, Poincaré had proved that if $r_0, \ldots, r_{n-1}$ were n regions with the same volume v included in a common region with volume V, and if $nv > kV$, then it was necessary that there were at least $k+1$ regions whose intersection was nonempty. Indeed, if one supposes that all the intersections taken $k+1$ by $k+1$ were empty, one may write (in modern notation) that $\sum_{i=0}^{n-1} \mathbb{1}_{r_i} \leq k$, hence $nv \leq kV$ by integrating over the volume V.

Let us still suppose valid the hypothesis of the previous theorem which asserts that the mobile point remains in a bounded region, in a portion of the space with volume V, and let us again take the discrete step τ in time. Let us next choose n sufficiently large so that $n > \frac{kV}{v}$. One may then find, among the n successive consequents of a region r_0 with volume v, $k+1$ consequents, denoted

$$r_{\alpha_0}, r_{\alpha_1}, \ldots, r_{\alpha_k}$$

with $\alpha_1 < \alpha_2 < \cdots < \alpha_k$, having a nonempty intersection denoted by s_{α_k}. Let us now call s_0 the α_kth antecedent of s_{α_k} and s_p, the pth consequent of s_0. If a mobile point starts from s_0, it will enter the regions

$$s_0, s_{\alpha_k - \alpha_{k-1}}, s_{\alpha_k - \alpha_{k-2}}, \ldots, s_{\alpha_k - \alpha_2}, s_{\alpha_k - \alpha_1}, s_{\alpha_k - \alpha_0}$$

which, by construction, are all included in r_0 (as for each $0 \leq i \leq k$, the α_ith consequent of $s_{\alpha_k - \alpha_i}$ is in s_{α_k} and therefore in r_{α_i}) . One has therefore shown that there are, in the considered region r_0, initial conditions of trajectories which pass at least $k+1$ times through r_0.

Let us eventually fix a region r_0 with volume v. Let us consider, writes Poincaré, σ_0 the subset of r_0 such that the trajectories issued from σ_0 do not pass through r_0 at least $k+1$ times between time 0 and time $(n-1)\tau$; denote by w the volume of σ_0. The probability p_k of the set of such trajectories is therefore $\frac{w}{v}$.

By hypothesis a trajectory starting from σ_0 does not pass $k+1$ times through r_0 , and hence not through σ_0. From the previous result, one has necessarily that

$nw < kV$, such that

$$p_k < \frac{kV}{nv}.$$

No matter how large k may be, one may choose n large so that this probability can be made as small as wanted. Poincaré, tacitly using the continuity of probability along a non-increasing sequence of events, concludes that the probability of the trajectories issued from r_0 which do not pass through r_0 more than k times between times 0 and ∞ is zero.

2.2. Kinetic theory of gases

As observed in the introduction, in 1892 Poincaré was not favorably disposed towards the statistical description of thermodynamics. His polemics with Tait, from which I quoted several passages, was closely tied to the mechanist spirit in which Poincaré had been educated. Statistical mechanics, and in particular the kinetic theory of gases, could not therefore pretend to be more than an ingenious construction with no explanatory value. An important text revealing Poincaré's thoughts on the subject was published immediately afterwards, in 1893, in one of the first issues of the *Revue de Métaphysique et de Morale* [66]. With great honesty, Poincaré mentioned the classical mechanical conception of the universe due to Newton and Laplace, but also the numerous problems it encounters when it tries to explain numerous practical situations of irreversibility, as in the case of molecular motion in thermodynamics. Poincaré mentioned that the kinetic theory of gases proposed by the English is the attempt *la plus sérieuse de conciliation entre le mécanisme et l'expérience* ([66], p. 536). Nevertheless, he stated that numerous difficulties still remained, in particular for reconciling the recurrence of mechanical systems (*"un théorème facile à établir"*, wrote the author, who may have adopted a humouristic posture) with the experimental observation of convergence towards a stable state. The manner in which the kinetic theory of gases pretends to evacuate the problem, by invoking that what is called a stable equilibrium is in fact a transitory state in which the system remains an enormous time, did not seem to convince our hero. However, at least, the tone adopted in [66] is obviously calmer than that of the exchanges with Tait. Another point which can be observed is that here, as in other works by Poincaré we shall comment on, Boltzmann was nowhere to be found. This absence, difficult to imagine as involuntary, remains unexplained, including for Von Plato in [90], p. 84.

In 1892, Lord Kelvin presented a note [47] to the Royal Society (of which he was then the president) with an unambiguous title. The note presented an *ad hoc* example demonstrating, in a supposedly *decisive* way, the failure of equipartition of kinetic energy following Maxwell and Boltzmann's theory. The two physicists had indeed deduced the equipartition of kinetic energy as a basic principle of their theory: the average kinetic energies of several independent parts of a system are in the same ratio as the ratio of the number of degrees of freedom they have. This result was fundamental for the establishment of a relation between kinetic energy and temperature.

In his short paper, Kelvin imagined a mechanical system including three points A, B, C, which are in motion in this order on a line KL, such that B remains almost motionless and only reacts to the shocks produced by A and C on one side and the other, whereas the mechanical situation on both sides is different because of a repulsive force F acting on A and pushing it towards B (in the zone KH of the scheme) while C can move freely.

FIGURE 1. Kelvin's construction in [47].

The total energy of C is balanced by the energy of A, but, as the latter includes a non-negative potential-energy term due to the repulsive force, Kelvin triumphantly concluded that the average kinetic energy of A and C cannot be equal, as they should have been following Maxwell's theory as the two points each have one degree of freedom. Kelvin commented:

> 'It is in truth only for an approximately "perfect" gas, that is to say, an assemblage of molecules in which each molecule moves for comparatively long times in lines very approximately straight, and experiences changes of velocity and direction in comparatively very short times of collision,

and it is only for the kinetic energy of the translatory motions of the molecules of the "perfect gas" that the temperature is equal to the average kinetic energy per molecule.' ([47], p. 399)

Reading this note encouraged Poincaré, as he noted, to reflect on the kinetic theory of gases, to understand whether Kelvin's objection was well founded, and to draw his own conclusions on the subject. At this precise moment, when he had been attacked by Tait, the title of the note by such an authority as Kelvin had impressed him and he may have thought he would find there a decisive argument confirming his own skepticism. And so in 1894 Poincaré published his first paper on the kinetic theory of gases [67]. Poincaré began by presenting a long general exposition of the fundamentals of Maxwell's theory. This survey seemed necessary in the first place because the kinetic theory of gases had been much less studied by French physicists than by English physicists[5]. These fundamentals were in the first place the ergodic principle, called the *postulate of Maxwell* by Poincaré, which asserts that, whatever may be the initial situation of the system, it will always pass an infinity of times as close as desired to any position compatible with the integrals of the motion; from this postulate Maxwell drew a theorem whose main consequence was precisely the point contested by Kelvin: in a system for which the only integral is the conservation of the kinetic energy, if the system is made up of two independent parts, the long run mean values of the kinetic energy of these two parts are in the same ratio as their numbers of degrees of freedom.

Poincaré began by observing, as previously in [66], that the recurrence theorem of [62] contradicted Maxwell's postulate along the recurrent solutions. It was therefore necessary at least to add that the postulate was true except for certain initial conditions.[6] As von Plato comments [90] (p. 84), we have here the formulation usually given today for the ergodic principle, in order to take into account the possibility of exceptional initial conditions. Once again, although this idea was also present in Boltzmann's works, the Austrian scientist was nowhere mentioned.

But it was above all the objection contained in Kelvin's paper that Poincaré desired to analyze in detail in order to check whether it contradicted Maxwell's results or not. From the situation of the system A, C, Poincaré built a representative geometric model: a point M in a phase space with three dimensions whose first coordinate is the speed of A, the second the speed of C and the third the abscissa of A. Using Kelvin's system conditions, he could define S, a solid of revolution, from which M cannot exit in the course of time. Naturally, two small regions included in S with the same volume can be entered a different number of times with the same total sojourn time because the speed in these volumes could be different. Poincaré introduced the notion of the *density of the trajectory* in a small element in S with volume v as the quotient $\frac{t}{v}$, where t is the total time spent by the trajectory in v ([67], p. 519). Using this representation, Poincaré could define the average value of the kinetic energy for A as the moment of inertia of S with respect to the plane

[5]'[...] a été beaucoup moins cultivée par les physiciens français que par les anglais.' ([67], p. 513)
[6]'[...] sauf pour certaines conditions initiales exceptionnelles' ([67], p. 518)

yz, for C as the moment of inertia with respect to xz, the 'masses' in S being distributed by the previously defined density. The solid S being one of revolution, these moments of inertia are equal: the fine analysis made by Poincaré therefore shows that one can recover the equipartition result by taking the average of the kinetic energies not uniformly over time but taking into account the phases of the motion and their duration.

Poincaré concluded his paper with a comment that may seem paradoxical in light of the result he had just obtained. While disputing the decisive character of Kelvin's arguments, Poincaré insisted that he nevertheless shared his colleague's skepticism. To give weight to his comment, he slightly transformed Kelvin's example in order to produce an *ad hoc* situation for which there *really is* a problem. In fact, some lines earlier, Poincaré had emphasized what was for him the fundamental point:

> 'I believe Maxwell's theorem really is a necessary consequence of his postulate, as soon as one admits the existence of a mean state; but the postulate itself must admit many exceptions.'[7]

For Poincaré, it was therefore the definition of the average states that posed a problem, and it was the search for a satisfactory definition that required attention from those those who wished to consolidate the bases of statistical mechanics. We shall see later that it was indeed in this direction that Poincaré, and later Borel, focussed their efforts.

2.3. Limit theorems

The textbook [69] published in 1896 constitutes the first of Poincaré's works dealing explicitly with the theory of probabilities. It was, as already mentioned, the result of lectures offered by Poincaré during the academic year 1893–94 at the Sorbonne, redacted by a former student of the École Normale Supérieure, who became an actuary and who probably wished to learn with the master; it was published by Georges Carré. This first edition does not have a preface, and presents itself as a succession of 22 lectures, more or less connected to each other, probably reflecting Poincaré's actual lectures. It is 274 pages long, compared to the 341 pages of the second edition of 1912 [77], which provides an idea of the number of complements. In its initial form, Poincaré's book appears as a successor to Bertrand's treatise [9], the framework of which it follows. Nevertheless, these textbooks really have a different feel, and we must leave it at that for the moment. However, as Poincaré's textbook has been discussed by several commentators, especially in [84] and [25], I shall restrict myself to just a few remarks. Let us note that the authors just mentioned focused on the 1912 edition, which naturally benefited from reflections of Poincaré after that very important period of the mid-1890s, when he was beginning to investigate the mathematics of randomness, so that this may not accurately

[7]'Je crois que le théorème de Maxwell est bien une conséquence nécessaire de son postulat, du moment qu'on admet l'existence d'un état moyen; mais le postulat lui-même doit comporter de nombreuses exceptions.' ([67], p. 521)

reflect the mathematician's state of mind in 1896. I choose here, in contrast, to focus on the original 1896 version.

An important part of Poincaré's textbook is devoted to the use of probability theory as a model of measurement error in the experimental sciences. In a commentary on his own works ([78], p. 121), Poincaré wrote:

> 'The Mathematical Physics Chair has for its official title: Calculus of Probability and Mathematical Physics. This connection can be justified by the applications of this calculus in all physics experiments, or by those in kinetic gas theory. In any case, for one semester I dealt with probability and my lectures were published. The theory of errors was naturally my main goal. I had to erect explicit reservations concerning the generality of the 'law of errors', but I sought to justify the law via new considerations in the cases where it remains legitimate.'[8]

In [69], the analysis of the law of errors begins on page 147 and occupies part of the following chapters. Poincaré commented on the manner in which the Gaussian character of the error had been obtained:

> '[This distribution] cannot be obtained by rigorous deductions. Several of its putative proofs are awful, including, among others, the one based on the statement that gap probability is proportional to the gaps. Nonetheless, everyone believes it, as M. Lippmann told me one day, because experimenters imagine it to be a mathematical theorem, while mathematicians imagine it to be an experimental fact.'[9]

Much of Poincaré's treatment of the law of errors, in particular the convergence to a Gaussian distribution of a Bayesian approach of the measurement process, was in fact already found by Laplace, at least in a preliminary form. Once again, it is not really possible to make precise what Poincaré knew. For details on what is today known as the Bernstein–von Mises theorem see [88] (p.140 *et seq.* and its references).

Let us consider observations of a phenomenon denoted by $x_1, x_2, \ldots, x_n$. The true measure of the phenomenon under study being z, the *a priori* probability that each of these n observations belong to the interval $[x_i, x_i + dx_i]$ is taken under the

[8]'La Chaire de Physique Mathématique a pour titre officiel : Calcul des Probabilités et Physique Mathématique. Ce rattachement peut se justifier par les applications que peut avoir ce calcul dans toutes les expériences de Physique; ou par celles qu'il a trouvées dans la théorie cinétique des gaz. Quoi qu'il en soit, je me suis occupé des probabilités pendant un semestre et mes leçons ont été publiées. La théorie des erreurs était naturellement mon principal but. J'ai dû faire d'expresses réserves sur la généralité de la "loi des erreurs"; mais j'ai cherché à la justifier, dans les cas où elle reste légitime, par des considérations nouvelles.'

[9]'[Cette loi] ne s'obtient pas par des déductions rigoureuses; plus d'une démonstration qu'on a voulu en donner est grossière, entre autres celle qui s'appuie sur l'affirmation que la probabilité des écarts est proportionnelle aux écarts. Tout le monde y croit cependant, me disait un jour M. Lippmann, car les expérimentateurs s'imaginent que c'est un théorème de mathématiques, et les mathématiciens que c'est un fait expérimental.'([69], p. 149)

form

$$\varphi(x_1, z)\varphi(x_2, z) \dots \varphi(x_n, z)dx_1dx_2 \dots dx_n.$$

Let finally $\psi(z)dz$ be the *a priori* probability so that the true value belongs to the interval $[z, z + dz[$.

Supposing that ψ is constant and that $\varphi(x_i, z)$ can be written under the form $\varphi(z - x_i)$, Gauss obtained the Gaussian distribution by looking for the φ such that the most probable value was the empirical mean

$$\overline{x} = \frac{x_1 + \dots + x_n}{n}.$$

Poincaré recalled ([69], p. 152) Bertrand's objections to Gauss' result; Bertrand had in particular disputed the requirement that the mean be the most probable value while the natural condition would have been to require it to be the *probable value* (which is to say the expectation).

Poincaré thus considered the possibility of suppressing some of Gauss' conditions. Keeping firstly the hypothesis that the empirical mean be the most probable value ([69], p. 155 – see also the details in [84], p. 149 *et seq.*), he obtained for the form of the error function

$$\varphi(x_1, z) = \theta(x_1)e^{A(z)x_1 + B(z)},$$

where θ and A are two arbitrary functions, B being such that the following differential equation is satisfied: $A'(z)z + B'(z) = 0$.

Considering next Bertrand's objection, Poincaré looked next at the problem that arises when one replaces the requirement of the *most probable value* with that of the *probable value*. Here ([69], p. 158) he gives a theorem he would use subsequently a number of times: if φ_1 and φ_2 are two continuous functions, the quotient

$$\frac{\int \varphi_1(z)\Phi^p(z)dz}{\int \varphi_2(z)\Phi^p(z)dz}$$

tends, when $p \to +\infty$, towards

$$\frac{\varphi_1(z_0)}{\varphi_2(z_0)},$$

where z_0 is a point in which Φ attains its unique maximum. Following his habit, which drove Mittag-Leffler to dispair, Poincaré's writing was somewhat laconic; he did not specify any precise hypothesis, or supply a real proof, presenting the result only as an extrapolation of the discrete case.

In any case, considering next

$$\Phi(x_1, \dots, x_n; z) = \varphi(x_1, z)\varphi(x_2, z) \dots \varphi(x_n, z),$$

Poincaré made the hypothesis that p observations resulted in the value x_1, p resulted in x_2, ..., p resulted in x_n, where p is a fixed and very large integer ([69], p. 157).

The condition requiring the mean to be equal to the expectation can therefore be written as

$$\frac{\int_{+\infty}^{+\infty} z\psi(z)\Phi^p(x_1,\ldots,x_n;z)dz}{\int_{+\infty}^{+\infty} \psi(z)\Phi^p(x_1,\ldots,x_n;z)dz} = \frac{x_1+\cdots+x_n}{n}.$$

Applying the previous theorem, under the hypothesis that Φ has a unique maximum in z_0, one has z_0 as the limit of the left-hand side, which must therefore be equal to the arithmetic mean $\overline{x}$. One is thus brought back to the previous question under the hypothesis that Φ should be maximal at $\overline{x}$. Under the hypothesis that φ depends only on the discrepancies $z - x_i$, Poincaré again obtained the Gaussian distribution. It is remarkable that the form of the *a priori* probability of the phenomenon ψ is not present in the result. This lack of dependence on the initial hypothesis might perhaps have been the inspiration for his method of arbitrary functions, described later.

Poincaré examined next the general problem by suppressing the constraint that φ depends only on the discrepancies, and obtained the following form for φ

$$\varphi(x_1, z) = \theta(x_1)e^{-\int \psi(z)(z-x_1)dz}$$

where $\int \psi(z)(z - x_1)dz$ is the primitive of $\psi(z)(z - x_1)$ equal to 0 in x_1.

He argued ([69], p. 165) that the only reasonable hypothesis was to take $\psi = 1$ as there was no reason to believe that the function φ, which depends on the observer's skillfulness, would depend on ψ, the *a priori* probability for the value of the measured quantity. For θ, on the contrary, there was no good reason to suppose it constant (in which case the Gaussian distribution would be again obtained). Poincaré took the example of the meridian observations in astronomy where a *decimal error* had been detected in practice: the observers show a kind of predilection for certain decimals in the approximations.

Poincaré gave a somewhat intricate justification for focusing on the mean because it satisfies a practical aspect: as the errors are small, to estimate $f(z)$ by the mean of the $f(x_i)$ was the same as estimating z by the mean of the x_i, as immediately seen by replacing $f(x)$ by its finite Taylor expansion in z,

$$f(z) + (x - z)f'(z).$$

In any case, the major justification was given in the following chapter (Quatorzième leçon, [69], p. 167), where the consistency of the estimator $\overline{x}$ was studied using an arbitrary law for the error, based on the law of large numbers. After having recalled the computation of the moments for the Gaussian distribution, Poincaré implemented the method of moments in the following way. Suppose that y, with distribution φ, admits the same moments as the Gaussian distribution. One then computes the probable value of $e^{-n(y_0-y)^2}$, where n is a given integer. Decomposing

$$e^{-n(y_0-y)^2} = \sum_{p=0}^{\infty} A_p y^{2p},$$

one obtains

$$\int_{-\infty}^{+\infty} \sqrt{h/\pi} e^{-hy^2} e^{-n(y_0-y)^2} dy = \sum_{p=0}^{\infty} A_p \mathbb{E}(y^{2p}),$$

letting $\mathbb{E}(y^{2p})$ denote the expectation of y^{2p} (the odd moments are naturally equal to zero), and h a positive constant. The same decomposition is valid by hypothesis if φ replaces the Gaussian distribution in the integral. One has therefore

$$\frac{\int_{-\infty}^{+\infty} \sqrt{h/\pi} e^{-hy^2} e^{-n(y_0-y)^2} dy}{\int_{-\infty}^{+\infty} \varphi(y) e^{-n(y_0-y)^2} dy} = 1,$$

and, using his theorem on the limits again, Poincaré could obtain, letting n tend to infinity,

$$\sqrt{h/\pi} e^{-hy_0^2} = \varphi(y_0).$$

Now, let us consider again that n measures

$$x_1, \ldots, x_n$$

of a quantity z are effectuated, and let us denote by $y_i = z - x_i$ the individual error of the ith measure. Let us suppose that the distribution of an individual error is arbitrary.

Poincaré began by justifying the fact of considering the mean

$$\frac{y_1 + \cdots + y_n}{n}$$

of the n individual errors as error. Indeed, he explained that the mean becomes more and more probable as the probable value of its square is

$$\frac{1}{n}\mathbb{E}(y_1^2),$$

and so, when n becomes large, the probable value of

$$\left(\frac{y_1 + \cdots + y_n}{n}\right)^2$$

tends towards 0 in the sense that the expectation

$$\mathbb{E}[(z - \overline{x})^2]$$

tends towards 0 (this is the L^2 version of the law of large numbers). As Sheynin observes ([84], p. 151), Poincaré made a mistake when he attributed to Gauss this observation as the latter had never been interested in the asymptotic study of the error. Poincaré then used his method of moments in order to prove that the distribution of the mean is Gaussian when the individual errors are centered and do not have a significant effect on it.

In the 1912 edition [77], Poincaré significantly added a section ([77], n°144 pp. 206–208) devoted to a proof of the central limit theorem and obtained a *justification* a posteriori *de la loi de Gauss fondée sur le théorème de Bernoulli.* Poincaré introduced the characteristic function as

$$f(\alpha) = \sum_x p_x e^{\alpha x}$$

in the *finite* discrete case, and

$$f(\alpha) = \int \varphi(x) e^{\alpha x} dx$$

in the case of a continuous density. In his mind, α was a real or a complex number and neither the bounds of the sum or the integral, nor the issue of convergence were mentioned. Poincaré's considerations on the Fourier and Laplace transforms can in fact already be largely found in Laplace, though the denomination *characteristic function* is Poincaré's. The Fourier inversion formula was used by Poincaré since his lectures on the analytical theory of heat [68] (see in particular the Chapter 6, p. 97). As usual, there is no mention of Laplace in Poincaré's text. Plausibly, Laplace introduced characteristic functions for probability distributions after having studied Fourier's treatise (see [22]). In the absence of proof that Poincaré knew Laplace's method, one may speculate that he had the same kind of illumination as his predecessor.

Thanks to Fourier's inversion formula, Poincaré stated that the characteristic function determined the distribution. He could thus obtain simply that a sum of independent Gaussian variables followed a Gaussian distribution and, by means of a heuristic and again quite laconic proof, that the error resulting from a large number of very small and independent partial errors[10] was Gaussian. It seems difficult to award the status of a proof of the central limit theorem to these few lines, a proof published some ten years later in works by Lindeberg and Lévy ([51], [50]). Besides, in this intriguing but rather hasty complement, Poincaré showed his complete ignorance of the Russian research on limit theorems (Čebyčev, Markov et Lyapunov) which produced some well established versions of the theorem.

In the sixteenth chapter of [69], (n°147 of the second edition [77] p. 211), Poincaré the physicist still had reservations about what would be an indiscriminate use of the theories he had just described, which depended so heavily on a mathematical idealization (the absence of systematic errors, overly smooth hypotheses...). He wrote, not without irony: 'I've argued the best I could up to now in favor of Gauss' law.'[11] He then focused on the study of exceptional cases and completed his textbook wit a detailed examination of the method of least squares; on these subjects, I refer the interested reader to the already quoted paper by Sheynin ([84]).

2.4. The great invention: the method of arbitrary functions

Although Poincaré is rightly considered to be the father of conventionalism in scientific philosophy, it would be simplistic to think that this position covers all

[10]'[...] résultante d'un très grand nombre d'erreurs partielles très petites et indépendantes.' ([77], p. 208)

[11]'J'ai plaidé de mon mieux jusqu'ici en faveur de la loi de Gauss.'

Poincaré's philosophy of research. Admittedly, from the very beginning, the latter had repeated that any use of probability must be based on the choice of a convention one had to justify. Thus, if one throws a die, one is generally led to take as a convention the attribution of a probability of 1/6 for each face to appear. However all the arguments used to justify this convention do not have the same value, and choosing well among them is also part of a sound scientific process. We cannot indeed be satisfied with common sense: Bertrand amused himself, when he constructed his famous paradoxes about the choice of a chord in a circle, in showing that the result depended so closely on the chosen convention that it lost meaning, and that the calculus of probability in such situations was reduced to a more or less ingenious arithmetic. The risk was to condemn the calculus of probability altogether as a vain science and conclude that our *instinct obscur* had deceived us ([70], p. 262).

And yet, wrote Poincaré, without this obscure instinct science would be impossible. How can one reconcile the irreconcilable?

Until then, common practice was to cite Laplace and use the principle of insufficient reason as a supporting argument. A dubious argument in fact, as in practice it amounts to assigning a value to the probability only by supposing that the different possible cases are equally probable, since we do not have any reason to assert the contrary. How could a scientist such as Poincaré, who was looking for a reasonably sound basis for using the mathematics of randomness, be satisfied with such a vicious circle?

Let us observe in passing that he was far from the first to deal with such a question. And besides, we have already seen that after Laplace's death, several weaknesses of his approach had been underlined: the vicious circle of the definition of probability by possibility, the absence of an answer to the general question of the nature of the probabilities of causes when applying Bayes's principle, let alone the confusions in ill-considered applications, in particular the judicial ones that we have already mentioned.... A substitute was sought for Laplace's theory. This problem of defining the *natural* value of probabilities had in particular obsessed German psychologists and physiologists throughout the second half of the 19th century ([46]). Von Kries in particular succeeded, a good ten years before Poincaré, in constructing the foundations of a method allowing one to justify the attribution of equal probabilities to the different outcomes of a random experiment repeated a large number of times ([45]). Poincaré, without question, completely ignored these works, all the more because they did not belong *stricto sensu* to the sphere of mathematics.

The question thus for Poincaré was to show that in some important cases, one may consider that the equiprobability of the issues in a random experiment as the result not only of common sense but also of mathematical reasoning, and thereby to avoid the criticism of Laplace's principle.

The idea developed by Poincaré, as earlier by von Kries, was that the repetition of the experiment a large number of times ends in a kind of asymptotic equilibrium, in a compensation, so that the hypothesis of equiprobability becomes reasonable even if one absolutely ignores the initial situation.

As early as the 1780s, Laplace observed that in many cases the initial distribution asymptotically vanishes when one repeatedly applies Bayes' method. For complements, the interested reader may consult Stigler's translation and commentary of Laplace ([86]), and also [20], p. 144. At the end of his paper [63], Poincaré briefly mentioned that the convention he had adopted to define the probability concept in the study of the stability of the three-body problem was by no means necessary: the result (a probability equal to zero for non-recurrent trajectories) remains true whatever the convention. Poincaré's aim in subsequent papers was to show that in more general situations, the knowledge of the probability distribution at the origin of time is not needed, as this distribution is not present in the final result. Two examples serve repeatedly as an illustration in the various texts ([70], [75] especially) in which he discussed his method: the uniform distribution of the so-called small planets on the Zodiac and the probabilities for the red and black slots of a roulette wheel. In [70], p. 266, Poincaré observed the proximity of these two situations. The systematic exposition of *arbitrary functions* as a fundamental method of the theory of probability seems due to Borel ([15], p. 114) but the expression *méthode des fonctions arbitraires* was later generally used in the context of Markov chains (see in particular [35]).

Let us first follow Poincaré's comments on the second, and simpler case of roulette ([70], p. 267). The ball, thrown with force, stops after having turned many times around the face of a roulette wheel regularly divided into black and red sectors. How can we estimate the probability that it stops in a red sector?

Poincaré's idea is that, when the ball travels for a large number of turns before stopping, any infinitesimal variation in the initial impulsion can produce a change in the color of the sector where the ball stops. Therefore, the situation becomes the same as considering that the face of the game is divided into a large number of red and black sectors. I make, said Poincaré, the convention that the probability for this angle to be fall between θ and $\theta + d\theta$ equals $\varphi(\theta)d\theta$, where φ is a function about which I do not know anything (as it depends on the way the ball moved at the origin of time, an *arbitrary* function). Poincaré nevertheless asserts, without any real justification, that we are *naturally* led to suppose φ is continuous. The probability that the ball stops in a red sector is the integral of φ estimated on the red sectors.

Let us denote by ε the length of a sector on the circumference, and let us consider a double interval with length 2ε containing a red and a black sector. Let then M and m be respectively the maximum and the minimum of φ on the considered double interval. As we can suppose that ε is very small, the difference $M - m$ is very small. And as the difference between the integral on the red sectors and the integral on the black sectors is dominated by

$$\sum_{k=1}^{\pi/\varepsilon} (M_k - m_k)\varepsilon,$$

(where M_k and m_k are respectively the maximum and the minimum on each double interval k of the subdivision of the face with length 2ε), this difference is small and it is thus reasonable to suppose that both integrals, whose sum equals 1, are equal to 1/2.

Once again, Poincaré's writing is somewhat sloppy. He emphasized the importance of the fact that ε was small with respect to the total swept angle,[12] but without giving much detail on how to interpret this fact. His brevity probably came from the parallel with the other example we shall now present – but which he studied earlier in his text.

The expression *small planets* designates the asteroïd belt present between Mars and Jupiter which had been gradually explored until the end of the 19th century. The first appearance of questions of a statistical type about these planets seems to go back to the twelfth chapter of the 1896 textbook ([69], p. 142), where Poincaré asked how one can estimate the probable value of their number N. For that purpose, he implemented a Bayesian method using the *a priori* probability for an existing small planet to have been observed, this probability being supposed to have a density f. It allowed him to carry out the computation of the *a posteriori* expectation of N.

In [70], Poincaré was interested in a remarkable phenomenon: the almost uniform distribution of the small planets in the different directions of the Zodiac. Poincaré looked for arguments justifying this fact ([70], p. 265 *et seq.*). We know, said Poincaré, that the small planets follow Kepler's laws, but on the contrary we absolutely ignore their initial distribution.

Let then b be the longitude of a small planet at the initial time, and a its mean motion. At time t, its longitude is therefore $at + b$. As already noted, one knows nothing of the initial distribution and we suppose it is given by an arbitrary function $\varphi(a, b)$, once more assumed regular in some way: Poincaré wrote that it was *continuous* but in the sequel used it as a function of class C^∞.

The mean value of $\sin(at + b)$ is given by

$$\int\int \varphi(a, b) \sin(at + b) da db.$$

When t becomes large, this integral becomes close to 0. Poincaré used successive integrations by part, with the derivatives of φ, whereas he could have used only continuity and the Riemann–Lebesgue lemma, but, as we have already seen, Poincaré did not regard the refinement of his hypotheses as a major concern. *A fortiori*, for every nonzero integer n, the integrals

$$\int\int \varphi(a, b) \sin n(at + b)\, da db, \quad \text{and} \quad \int\int \varphi(a, b) \cos n(at + b)\, da db$$

[12]'[...] par rapport à l'angle total parcouru.'

are also very small for a large fixed t. Therefore, if one denotes by ψ the probability density of the longitude at time t, one has for every $n \geq 1$,

$$\int_{[0,2\pi[} \psi(u) \sin nu \, du, \quad \text{and} \quad \int_{[0,2\pi[} \psi(u) \cos nu \, du$$

very close to 0. The Fourier expansion of ψ leads to the conclusion that ψ is almost constant, that is to say, that the longitude of a small planet is roughly uniformly distributed on the Zodiac.

2.5. Cards shuffling

If the example of the small planets illustrates sensitivity to initial conditions, that of kinetic gas theory is connected with the complexity of causes. The number of molecules is so large, and they collide in so many ways, that it is impossible to consider the system they form as simply describable by classical mechanics. In 1902 the first textbook was published expounding the basic principles of statistical mechanics, written by Gibbs ([37]). It developed two main applications for the new theory: in addition to the kinetic theory of gases, it introduced the situation of mixing two liquids (a drop of ink put in a glass of water) in order to present the evolution of a system towards equilibrium. Hadamard, in 1906, had written a review of Gibb's book for the *Bulletin des sciences mathématiques* ([38]). In order to illustrate this mixing situation, he invented the ingenious metaphor of the shuffling of a pack of cards by a gambler evolving towards an equal distribution of the possible permutations of the cards. Hadamard however did not propose any mathematical treatment of the question and it was Poincaré, in the paper he published in 1907 in Borel's journal ([75] republished later in [76] and [77]), which first analyzed the problem. He restricted himself in fact to the simplest case, that of two cards. Let us suppose, said Poincaré, that one has a probability p that after one permutation, the cards are still in the same order as before the permutation, and $q = 1 - p$ that their order is reversed. Let us consider there are n successive permutations and that the gambler who shuffles the cards earns a payoff S equal to 1 franc if the order after these n permutations is unchanged, and -1 franc if it is reversed. A direct computation of the expectation shows that

$$\mathbb{E}(S) = (p - q)^n,$$

as, in a modern formulation, S can be written as

$$\prod_{i=1}^{n} X_i$$

with the $X_i = \pm 1$ independent with distribution (p, q) representing the fact that the ith permutation has changed or not the order of the cards. Hence, except in the trivial cases $p = 0$ or 1, $\mathbb{E}(S) \to 0$ when n tends towards infinity, which amounts to saying that the two states $+1$ and -1, and therefore the two possible orders, tend to become equiprobable. It is interesting that Poincaré was inspired, for the recursive computation of the expectation without first looking for the distribution of S, by several computations of expectations he found in Chapter III of Bertrand's book [9].

As mentioned by Poincaré, the tendency to uniformity remained true whatever the number of cards but the *démonstration serait compliquée.* One may suppose that in 1907 Poincaré already had the idea of the proof for the general case of n cards but he wrote it only for the second edition of his textbook of probability in 1912, in the first section of a chapter added to the book, entitled *Questions diverses.* Curiously, Poincaré's method of proof, contrary to what he had done in the case of two cards, was not inspired by probabilistic reasoning but was connected to the theory of groups. Though the Perron–Frobenius theory was already available (but probably unknown to Poincaré), Poincaré referred to older works by Frobenius published in the *Sitzungsberichte* of Berlin Academy between 1896 and 1901, and by Elie Cartan [24] that he had himself extended in his paper [72] (consult [83] and [25] for details). We shall see in the next part that this non-probabilistic aspect did not escape Borel, who proposed an alternative approach. As for card shuffling, it enjoyed a spectacular renewed interest in the 1920s.

3. Third part: an uneven heritage

We now tackle the heritage of Poincaré's ideas about randomness and probability. This is an intricate question. Indeed, Poincaré cannot be considered as a full probabilist in the spirit of mathematicians of later generations like Paul Lévy and Andrei Nikolaevich Kolmogorov. As we have already mentioned, these studies on probability constitute only a very small island in the ocean of the mathematician's production. Moreover, it is rather difficult to locate a very precise result, a theorem, concerning the theory of probability which can be specifically credited to Poincaré. His primary goal was to refine already existing results or to explore new aspects and new questions while not feeling compelled to give them a complete structure. We should again repeat here that in this domain more than any other, it was above all Poincaré whom Poincaré sought to convince, and therefore his works dealing with probability, including his philosophical texts, often take on a rambling tone, written following his train of thought, often slightly verbose, illustrating Picard's opinion (as reported by K. Popoff): he ignored the adage *pauca sed matura*[13]. As Bru notes ([20], p. 155), everyone at that time had read Poincaré. But one has the impression that few understood what he wrote about probability.

3.1. Borelian path

Émile Borel was unquestionably the main exception. Not only did he read and understand Poincaré, but he was poised to make the subject his own in a spectacular way, so much so that he may be regarded as the first French probabilist of the twentieth century. We shall examine how this passing of the baton took place between the master and his young disciple.

[13] 'Il ne connaissait pas l'adage *pauca sed matura.*' ([80], p. 89)

It must firstly be said that this probabilistic turn of Émile Borel was one of the most singular changes one could observe in a mathematician around 1900. After initiating a profound transformation of methods in the theory of functions, Borel became a star of mathematical analysis in France. Nothing seemed to predispose him to take the plunge and to devote important efforts to study, refine and popularize the calculus of probability whose dubious reputation in the mathematical community – on which we commented above – might have led to rebuke. The context of Borel's turn since 1905, the date of publication of his first work in this domain, has been studied in detail several times, for instance in the papers [32] and [59]. The main difference which can be found between the discovery of probability by Poincaré and by Borel is that, for the latter, it arose from reflections within the mathematical field and more specifically by considerations on the status of mathematical objects – in particular about real numbers. In Borel, during the years just preceding 1900, we note indeed a greater and greater distance from Cantorian romanticism and its absolutist attitude, as emphasized by Anne-Marie Décaillot in her beautiful book on Cantor and France ([31], p. 159). Borel gradually replaced this idealistic vision, which no longer satisfied him, by a realism colored by a healthy dose of pragmatism: the probabilistic approach appeared then to Borel as an adequate means by which to confront various forms of reality: mathematical reality first and then physical reality and practical reality....

The best synthesis summing up Borel's spirit about the quantification of randomness can be found in Cavaillès' text [27] published in the *Revue de Métaphysique et de Morale*; it should be seen, at least in part, as a commentary on Borel's essay on the interpretation of probabilities [18] that completed the great enterprise of the *Traité du Calcul des Probabilités et de ses Applications* begun in 1922. As Cavaillès lyrically put it ([27], p. 154), probabilities appear to be the only privileged access to the path of the future in a world which is no longer equipped with the sharp edges of certainty, but presents itself instead as a hazy realm of approximations. Borel, at the moment of his probabilistic turn thirty years earlier, expressed himself similarly when he asserted that a coefficient of probability constituted the clearest answer to many questions, an answer which corresponded to an absolutely tangible reality, and when he was ironic about reluctant spirits who preferred certainty, and who would perhaps also prefer that 2 plus 2 make 5.[14]

I refer the reader to the aforementioned studies for precise details on these questions. What I would like to consider here is how Borel had combined his research on calculus of probability with the considerations of his predecessor. From his very first paper, Borel announced that he adopted the conventionalism of Poincaré ([13], p. 123). But his aim was to illustrate the role that the (then novel) Lebesgue integral and measure theory could play, after he discovered with amazement their use in [95] by the Swedish mathematician Anders Wiman (on this subject, see [32]).

[14] '[...] peut-être aussi que 2 et 2 fissent 5.'

> 'The methods adopted by Mr. Lebesgue allow us to examine [...] questions of probability that appear inaccessible to the classical methods of integration. Moreover, in the simpler cases, it suffices to use the theory of sets I called 'measurable', and which Mr. Lebesgue later termed 'measurable (B)'; the first use of this theory of measurable sets for the calculation of probability is due, I believe, to Mr. Wiman.'[15]

I shall not deal here with the radical transformations the Lebesgue integral brought to analysis at the beginning of the twentieth century. For a broad overview, one consult [41]. Nevertheless, for the sake of completeness, let us say at least a few words about Borel's role in the elaboration of this theory.

In his thesis dealing with questions of the extension of analytic functions, Borel invented a new concept of analytic extension, more general than that of Weierstrass, using a great deal of geometric imagination. In the course of his proof, he proved that a countable subset of an interval can be covered by a sequence of intervals with total length as small as one wants. This was probably the first appearance of a σ-additivity argument for the linear measure of sets. In subsequent years, Borel considerably fleshed out his construction, in particular in his work [12], by introducing the notion of measurable set and of measure based on σ-additivity. These concepts had however a limited extension for Borel as he considered only explicit sets obtained by countable unions and complementary sets, forcing him to make the shaky suggestion that one should attribute a measure *inferior to* α to any subset of a measurable set with measure α.

One had to wait for Lebesgue's thesis and the publication of his Note [49], in which he introduced a new conception for integration, for the notion of measurable set to reach its full power, on which is based the remarkable flexibility of the integral exploited by Borel in his paper [13]. There he showed in particular how the use of Lebesgue's integral can allow one to give meaning to some questions formulated in a probabilistic way. One of the most simple of these is that of assigning zero probability to the choice of a rational number when drawing a real number at random from the interval [0,1]. Let us insist on the fact that for Borel, it was more critical that the Lebesgue integral give a question meaning, than to provide its answer. We see here Borel being absolutely in line with Poincaré's conventionalism, but the choice of the convention (identifying the probability with the measure of a subset of [0,1]) is based on mathematical considerations.

However it is above all in his long paper of 1906 on the kinetic theory of gases ([14]) that Borel would fit in Poincaré's heritage, at the same time introducing new considerations showing that he was also striking out on his own. He often insisted on the difference between his approach and that of Poincaré ([14], p. 11, note 2).

---

[15]'Les méthodes de M. Lebesgue permettent d'étudier [...] des questions de probabilités qui paraissent inaccessibles par les procédés d'intégration classique. D'ailleurs, dans les cas particuliers les plus simples, il suffira de se servir de la théorie des ensembles que j'avais appelés mesurables et auxquels M. Lebesgue a donné le nom de mesurables (B); l'application de cette théorie des ensembles mesurables au calcul des probabilités a été, à ma connaissance, faite pour la première fois par M. Wiman.' ([13], p. 126)

Borel's aim was to provide a genuine mathematical model for Maxwell's theory in order to satisfy mathematicians.

> 'I would like to address all those who shared Bertrand's opinion about the kinetic theory of gases, that the problems of probability are similar to the problem of finding the captain's age when you know the height of the mainmast. If their scruples are somewhat justified in virtue of the fact that you cannot fault a mathematician for his love of rigor, nevertheless, it does not seem to me impossible to give them satisfaction. Such is the aim of the following pages: they do not advance the theory in any real sense from a physical point of view, but perhaps they will convince numerous mathematicians of its interest, and by increasing the number of investigators, will contribute indirectly to its development. Should this happen, these pages will not have been useless, independently of the aesthetic interest inherent to any logical construction.'[16]

Thus a motivation for Borel's agenda was that he regarded the various considerations of Poincaré on kinetic theory as insufficient to convince mathematicians of its interest. Let us observe in passing that Poincaré, that same year 1906, wrote a new paper for the *Journal de Physique*, where he studied the notion of entropy in the kinetic theory of gases ([74]); there was probably no direct link between Poincaré's and Borel's publications, which treat different questions.

Borel began his paper by returning to one of Poincaré's major themes: the distribution of the small planets. However, he approached it from a new angle (see below). He then applied the results he obtained to the construction of a mathematical model from which Maxwell's distribution law can be deduced. Borel's fundamental idea was that in the phase space where coordinates are the speeds of n molecules, the sum of squares of these speeds at a given time t is equal (or, more exactly, proportional) to n times the mean kinetic energy, so that the point representing the system of the speeds belongs to a sphere with a radius proportional to $\sqrt{n}$. Borel went on to perform an asymptotic study of the uniform measure on the ball with radius $\sqrt{n}$ in dimension n. I refer the interested reader to [90], [91] et [57] for details, and shall restrict myself to some comments on the first part of [14], dealing with the small planets.

Considering a circle on which there are points representing the longitudinal position of the small planets, Borel posed the following question: What is the

[16]'Je voudrais m'adresser à tous ceux qui, au sujet de la théorie cinétique des gaz, partagent l'opinion de Bertrand que les problèmes de probabilité sont semblables au problème de trouver l'âge du capitaine quand on connaît la hauteur du grand mât. Si leurs scrupules sont justifiés jusqu'à un certain point parce qu'on ne peut reprocher à un mathématicien son amour de la rigueur, il ne me semble cependant pas impossible de les contenter.'

'C'est le but des pages qui suivent : elles ne font faire aucun progrès réel à la théorie du point de vue physique; mais elles arriveront peut être à convaincre plusieurs mathématiciens de son intérêt, et, en augmentant le nombre de chercheurs, contribueront indirectement à son développement. Si c'est le cas, elles n'auront pas été inutiles, indépendamment de l'intérêt esthétique présent dans toute construction logique.' ([14], p. 10) probability for all the small planets to be situated on the same half-circle fixed in advance? As Borel noted, if one had perfect knowledge of the positions of the planets, there would be no reason to invoke probabilities, as one could directly assert whether the event was realized or not. He argued that it was necessary to restate the question to give it well-defined probabilistic meaning according to a selected convention. The simplest convention would be to assume that the probability for each planet on the chosen half-circle C_1 is equal to the probability of being on the complementary half-circle (and therefore equal to $1/2$), and that the different planets are situated independently with respect to each other. In this case, naturally, if there are n planets, the desired result is $1/2^n$. However, if this independence was more or less tacitly considered by Poincaré, Borel challenged it as being questionable, the planets having clearly mutual influences ([14], p. 12), and so he sought to forgo this hypothesis. Progressively enlarging his initial problem, Borel arrived at the following asymptotic formulation ([14], p. 15):

> Problem C. – Given the mean motions of n small planets to within ε and their exact initial positions, one denotes by ϖ the probability that, at a time t chosen at random in an interval a, b, every corresponding point P is in C_1. What is the limit to which ϖ tends when the interval a, b increases indefinitely?[17]

Borel could then implement a method of arbitrary functions in dimension n without supposing the initial independence of the motions of the planets, and prove that asymptotically the desired probability was $1/2^n$, a type of ergodic theorem which showed an asymptotic independence he would also show in the case of his model for the kinetic theory of gases having Gaussian distributions as limits. A rather curious detail is that the result given by Borel, proving the convergence of the uniform distribution on a sphere with radius $\sqrt{n}$ in n-dimensional space to independent Gaussian variables, is today called *Poincaré's Lemma*, even though it is entirely absent from the works of Poincaré (for complements about this strange fact, see [57] and the references therein).

I have not been able to make out clearly whether Poincaré was ever interested in the research and the work of his successor in the field of probability. The only sign which might indicate at least a passing interest is the fact that he agreed to write an article entitled "Le hasard" ([75]) for the *Revue du Mois*. But, to my knowledge, Poincaré did not comment on Borel's work, and what is even more surprising, if one remembers Poincaré's work at the beginning of the 1890s, he showed no interest in measure theory as applied to the mathematics of randomness. Poincaré, here again, stood poised on the threshold of a domain he helped to create, but did not enter.

---

[17] 'Problème C. – Connaissant à ε près les moyens mouvements des n petites planètes et connaissant exactement leurs positions initiales, on désigne par ϖ la probabilité pour qu'à une époque t choisie arbitrairement dans un intervalle a, b tous les points P correspondants soient sur C_1. Quelle est la limite vers laquelle tend ϖ lorsque l'intervalle a, b augmente indéfiniment?' ([14], p. 15)

3.2. Markovian descent

In order to complete this outline of Poincaré's probabilistic heritage, let us finally consider what may have been the most amazing consequence of his work: the dazzling development, since the end of the 1920s, of the theory of Markov chains and Markov processes. This story has already been set out in several texts and I shall again restrict myself to comments on only the most salient points, referring the reader elsewhere for more information.

We have already evoked Poincaré's investigations of card shuffling and the fact that in his proof of the convergence to the uniform distribution in [77], he used an algebraic method with limited exploitation of the probabilistic structure of the model. Borel, an attentive reader, immediately realized this and wrote a note, asking Poincaré to present it to the Academy for the *Comptes-Rendus de l'Académie des Sciences* (the only letter from Borel placed online on the website of the Archives Poincaré[18]).

Borel wrote to his colleague on 29 December 1911:

> 'I have just read the book you kindly sent me; I do not need to tell you how much the new sections interested me, in particular your theory of card shuffling. I tried to render it accessible to those unfamiliar with complex numbers, and it seems to me that in doing so I reached a slightly more general proposition. If it is new, and if you find it interesting, I would ask you to communicate the attached note.[19]

Poincaré acted immediately and the note was presented on 2 January 1912. Borel's method in [16] was in fact an extension of the elementary one used by Poincaré to treat the case of two cards, where one looks at the evolution of the successive means in the course of time. This method later became the standard in proofs of the exponential convergence of an irreducible finite Markov chain towards its stationary distribution (see for instance [10], p. 131). Here the stationary distribution is uniform due to the reversible character of the chain. Borel even gave himself the satisfaction of introducing a dependence on time (the chain becoming inhomogeneous).

He considered the regular case where there exists an ε such that, at every moment, the transition probabilities of one permutation to another at a subsequent time are all greater than some ε. In Borel's notation, let $p_{j,n}$ be the probability of the jth possible permutation of the cards before the nth operation. Denoting by $\alpha_{j,h,n}$ the probability for A_h to be replaced by A_j during the nth operation, one

---

[18] http://www.univ-nancy2.fr/poincare/chp/

[19] 'Je viens de lire le livre que vous avez eu l'amabilité de me faire envoyer; je n'ai pas besoin de vous dire combien les parties neuves m'ont intéressé, en particulier votre théorie du battage. J'ai essayé de la mettre à la portée de ceux qui ne sont pas familiers avec les nombres complexes, et il m'a semblé que j'obtenais ainsi une proposition un peu plus générale. Si elle est nouvelle, et si elle vous paraît intéressante, je vous demanderai de communiquer la note ci-jointe.'

has

$$p_{j,n+1} = \sum_{h=1}^{h=k} \alpha_{j,h,n} p_{h,n}$$

with the constraint $\sum_{h=1}^{k} \alpha_{j,h,n} = 1$ where k denotes the number of possible permutations. Let us immediately observe that P_n and p_n, the largest and the smallest of the $p_{j,n}$, form two sequences, respectively nonincreasing and nondecreasing. Let P and p denote their limits. For a given $\eta > 0$, one may choose n for which $P_n \leq P + \eta$, and therefore the $p_{j,n}$ are inferior to $P + \eta$. After N operations one can write

$$p_{j,n+N} = \sum_{h=1}^{h=k} \beta_{j,h,n} p_{h,n} \ , \ \sum_{h=1}^{h=k} \beta_{j,h,n} = 1$$

where the $\beta_.$ are the transition probabilities between time n and time $n+N$, each being greater than ε by hypothesis.

Let us consider the smallest of the $p_{h,n}$, $p_{h_0,n}$ so that $p_n = p_{h_0,n} \leq p$. For the sake of simplicity, let us denote by β its coefficient $\beta_{j,h_0,n}$; by hypothesis, $\beta \geq \varepsilon$. Let us observe that $\sum_{h=1, h\neq h_0}^{h=k} \beta_{j,h,n} = 1-\beta$. Therefore, one can write, by choosing j such that $p_{j,n+N}$ is superior or equal to P,

$$P \leq p_{j,n+N} \leq \beta p + (1-\beta)(P+\eta) = P + (1-\beta)\eta - \beta(P-p)$$

and hence

$$P - p \leq \frac{1-\beta}{\beta}\eta \leq \frac{1-\varepsilon}{\varepsilon}\eta.$$

η being arbitrary small, one concludes that $P = p$ and therefore that asymptotically the $p_{j,n}$ become all equal to $1/k$. Let us observe in passing that in these blessed years when it was permitted to publish mistakes, Borel erred in writing his inequality, considering ε instead of the number we called β, a fact which naturally did not change anything in the final result.

Nobody seemed to have paid attention to Borel's note: when these results were rediscovered by Lévy and then Hadamard in the 1920s, neither of them had the slightest idea of its existence (on this subject see the letters from Lévy to Fréchet [6], pp. 137 to 141).

We must next skip five years and cover several hundreds of kilometers east in order to see a new protagonist coming on stage, the Czech mathematician Bohuslav HOSTINSKÝ. Moreover, as if it were not enough that we must invoke an unknown mathematician, we must first say a few words about an unknown philosopher. Indeed the man who may have been, together with Borel, the most attentive contemporary reader of Poincaré's texts on probability was another Czech, the philosopher Karel VOROVKA (1879–1929), whose influence on Hostinský was decisive.

It is not possible here to discuss this singular figure in detail and I shall therefore restrict myself to giving some elements explaining how he had got involved in this melting pot. An interesting and very complete study on Vorovka was published in Czech some years ago [60] and hopefully it will become more accessible in a more widely known language. Some complements can also be found in [56] and the references therein. Two reasons explain this general ignorance of Vorovka: the fact that his works, mostly in Czech, were never translated, and also that his early death precluded the collection of his ideas in a large-scale work. Placing himself in the tradition of Bernhard Bolzano (1771–1848), the major figure of the philosophical scene in Prague during the 19th century, Vorovka sought an approach combining both his strong mathematical education and a rather strict religious philosophy, an original syncretism of empiricism and idealism which had close links with the thought of the hero of the Czechoslovak independence, T.G. Masaryk, and with American pragmatist philosophy, in which he had much been interested.

Vorovka's discovery of Poincaré's philosophical writings at the beginning of the 20th century was a real revelation: he drew from them the conviction that the scientific discoveries at the end of the 19th century, especially in physics, compelled a reconsideration of the question of free will. Vorovka showed a real originality in that he did not content himself with principles, but closely studied the mathematical problems raised by the theory of probability. He was a diligent reader of Bertrand's textbook, of Borel's texts, but also of Markov's works, publishing several works inspired by papers of the Russian mathematician (see [54], [92], [93], [94]). At the time when he was granted tenure at the Czech University in Prague, around the year 1910, Vorovka met the mathematician Bohuslav Hostinský, who had just returned to Bohemia after a period of research in Paris. In Hostinský's own words (see [44]), it is through the discussions he had with Vorovka that he learned about Poincaré's works, and he began to reflect upon the calculus of probability, a domain somewhat remote from his original field of research (differential geometry).

Following Jiří Beránek, who was one of the last assistants to Hostinský after the Second World War at the University of Brno, another source of the latter's interest in the calculus of probability is found in his reading of the 1911 paper by Paul and Tanya Ehrenfest on Statistical Mechanics for the *Encyklopädie der Mathematischen Wissenschaften*, translated and completed by Borel for the French version of the *Encyclopédie des Sciences Mathématiques* [33].

Beránek wrote ([8]) that this paper, whose impact was considerable,

> 'put the emphasis on statistical methods in physics, along with geometrical methods, mainly in connection with the works of L. Boltzmann on kinetic gas theory. Boltzmann's work sustained discussions and controversies concerning the correctness and legitimacy of his mathematical methods. Hostinský, as he mentioned later, began to study Boltzmann's works in 1915, and to take an interest in the efforts made to provide precise mathematical bases for the kinetic theory. The central point of these efforts implied a reexamination of some fundamental questions of the theory of probability. Hostinský was especially impressed by the fundamental works of H. Poincaré on the bases of probability calculus which opened the way for new methods necessary for the improvement

of kinetic theory. For this reason, around 1917, Hostinský began to study in earnest questions in the calculus of probability... '[20]

The fact that Hostinský began to deal seriously with probability in 1917 is attested by his own diary, kept in the archives of Masaryk's university in Brno. The diary's entries concern only comments on differential geometry until 1917. On 10 January 1917, Hostinský made some observations on the study of card shuffling by Poincaré, following [77], and on January 18th he took up problems of lottery. A first paper appeared some months later in the *Rozpravy České Akademie* dealing with the problem of Buffon's needle [43].

The problem of Buffon's needle is a classic of the calculus of probability and Hostinský began by expounding it:

> 'A cylindrical needle is thrown on a horizontal floor, on which are traced equidistant parallels; the distance $2a$ between two successive parallels is supposed larger than the length $2b$ of the needle. What is the probability that the needle meet one of the parallels?'[21]

Buffon had proposed a solution whose numerical result $\frac{2b}{\pi a}$, in which π was present, was a source of numerous propositions for an 'experimental' calculation of π. But, in fact, Buffon's proof was based on the hypothesis that the needle center could be located anywhere on the plane, and Hostinský, in a second critical part of his paper, mentioned the dubious nature of such a hypothesis, just as Carvallo had done before him in 1912. An experimental device could only take the form of a table of limited size, and it is then clear that, depending on the choice of a small square C_1 at the center of the table or another square C_2 on the edge of the table with the same area, the probability p_1 that the center of the needle belongs to C_1 and the probability p_2 that it belongs to C_2 cannot be the same: indeed, C_2 is strongly subject to the constraint that the needle does not fall from the table, but C_1 is very weakly so constrained, so that intuitively one should have $p_1 \gg p_2$.

Hostinský therefore considered it indispensable to suppose unknown the *a priori* distribution of the localization of the needle. It is an *unknown* distribution (with density) $f(x, y)dxdy$. But, mentioned Hostinský, Poincaré also, in the resolution of several problems of probability, allowed the use of such an *arbitrary*

[20]'[...] mettait l'accent sur les méthodes statistiques en physique, à côté des méthodes géométriques, principalement en relation avec les travaux de L. Boltzmann sur la théorie cinétique des gaz. Sur ceux-ci furent menées discussions et controverses, au sujet de l'exactitude et de la légitimité des méthodes mathématiques employées. Hostinský, comme il l'a lui même mentionné, commença à partir de 1915 à étudier les travaux de Boltzmann et à s'intéresser aux efforts qui étaient faits pour donner à la théorie cinétique des bases mathématiques précises. Le point central de ceux-ci nécessitait un nouvel examen de certaines questions fondamentales de la théorie des probabilités. Hostinský fut particulièrement impressionné à ce sujet par les travaux fondamentaux de H. Poincaré sur les fondements du calcul des probabilités qui ouvraient la voie à de nouvelles méthodes nécessaires pour le perfectionnement de la théorie cinétique. Pour cette raison, vers 1917, Hostinský commença à s'occuper sérieusement de questions de calcul des probabilités...'

[21]'On lance une aiguille cylindrique sur un plan horizontal, où sont tracées des parallèles équidistantes; la distance $2a$ de deux parallèles voisines est supposée plus grande que la longueur $2b$ de l'aiguille. Quelle est la probabilité pour que l'aiguille rencontre l'une des parallèles?'

density and observed that in some situations this function would not be present in the final result. Hostinský proposed to prove that if a domain A of the space is segmented in m elementary domains with the same volume ε, and containing each a white part with volume $\lambda\varepsilon$ and a black part with volume $(1-\lambda)\varepsilon$ (where $0 < \lambda < 1$), then for any sufficiently regular function $\varphi(x, y, z)$, the integral on the white parts will asymptotically (when m tends to infinity) be equal to λ times the integral of φ on A.

Hostinský then applied this result in order to propose a *new* solution to the problem of the needle. Instead of Buffon's unrealistic hypothesis, he supposed that the center of the needle is compelled to fall in a square with side $2na$, $n \in \mathbb{N}$, with a density of probability given by an unknown function φ (which he supposed to have bounded derivatives) and kept on the other hand the second hypothesis concerning the uniform distribution of the angle ω of the needle with respect to the parallels. This being set out, dividing the domain of integration $0 < x < 2na$, $0 < y < 2na$, $0 < \omega < \frac{\pi}{2}$ in n^2 subdomains (by partitioning the values of x and y with respect to the multiples of a), each small domain is itself divided into two parts (corresponding to the fact that the needle intersects [*white* part] or does not intersect [*black* part] the corresponding parallel). The ratio of their respective volumes to the total volume of the subdomain is constant and equal, for the white part, to $\frac{2b}{\pi a}$. An application of the previous theorem then allows one to assert that we obtain the desired probability, at least asymptotically when n tends towards infinity.

In the Spring of 1920, seeking to benefit from the sympathy of French public opinion towards the young Czechoslovakia, Hostinský had sent to Émile Picard the translation of his paper and Picard proposed immediately (18 April 1920) to include it in the Mélanges of the *Bulletin des Sciences Mathématiques*. This slightly revised version of the paper of 1917 was published at the end of 1920 and Maurice Fréchet, who had just arrived in Strasbourg and considered himself as a missionary [85] read it with attention, as he mentioned in a subsequent letter to Hostinský, dated 7 November 1920, in which he congratulated him on having obtained such a *positive result*.

As we have just explained, following Poincaré's example, Hostinský required that the function φ admit a uniformly bounded derivative in the domain A in order to obtain an upper bound for the difference between the maximum and the minimum of φ on each of the small domains. But Fréchet, when he read the paper, rightly realized that as only an estimation of the integrals of φ on these domains was needed, the simultaneous convergence of the superior and inferior Darboux sums towards the integral of φ allowed one to obtain the desired result with φ Riemann-integrable. This is what he wrote, together with the proof, to Hostinský on 7 November 1920.

It seems that the former letter refers to Fréchet's initial research on probabilistic questions. It was subsequently published in a short note in 1921 ([34]). Hostinský answered on 22 December 1920, agreeing with Fréchet that the integration hypothesis was sufficient. He also mentioned that Borel had already suggested that Poincaré's hypothesis could be weakened, supposing only the function to be continuous. In his textbook on probability published in 1909 by Hermann [15], in which Borel devoted the whole of Chapter VIII to the introduction of arbitrary functions by considering both Poincaré's examples of the roulette wheel and of the small planets on the Zodiac, Borel noted that the hypothesis of continuity was sufficient to apply Poincaré's method. Fréchet was to include Hostinský's observation in his note in 1921 [34] (where he emphasized that it had been inspired by the latter after having read his paper on Buffon's needle). In [34], he mentioned Borel's work to emphasize that hypotheses of both continuity (Borel) and derivability (Poincaré) were useless and that Riemann-integrability was sufficient.

His friendly relationship with Fréchet encouraged Hostinský to continue his probabilistic studies, and this leads us eventually to the last step of this long journey, introducing Jacques Hadamard (1865–1963). The presence of this name in our story may seem quite strange, and, in fact, Hadamard was interested in probabilities only during one semester of the academic year 1927–1928. He had never considered them before, and would never do so again, showing even some irritation towards Lévy, one of the disciples of whom he was most fond, when he 'wasted' his mathematical talent in the 1920s and left the royal path of functional analysis for the calculus of probability.

Following Poincaré's example, Hadamard always kept in mind physical theories from which he intended to extract new mathematical problems. It was from this perspective that he had written the aforementioned review of Gibbs' book in 1906.

When Hadamard began writing up his course of analysis at the École Polytechnique in the 1920s (published by Hermann in two volumes in 1926 and 1930), he had to prepare some lectures in 1927 on probability theory, and he took up Poincaré's example of card shuffling. On this occasion, he recovered Borel's method of successive means and published in 1927 a note in the *Comptes-Rendus de l'Académie des sciences de Paris* [39]. Soon after that, Hostinský discovered Hadamard's note and sent an extension of it for publication in the *Comptes-Rendus* [42], which appeared in the first weeks of 1928. There, for the first time, before everyone – except for Bachelier, but, alas, who had ever read Bachelier among the mathematicians I write about! – and especially before Kolmogorov, Hostinský introduced a Markovian model in continuous time. At the Bologna congress in September 1928, it was realized that Poincaré's card-shuffling studies were in fact a special case of the model of variables in chain introduced by Markov in 1906, and developed in several of his posterior publications as well as by Bernstein in 1926, but which were largely ignored outside Russia. The attention this drew, in particular at the Congress in Bologna in 1928, inaugurated intense activity on these questions which continued throughout the 1930s, a story brilliantly recounted in [20] to which I refer the interested reader. This unexpected crowning of Poincaré's efforts seems to be a perfect moment to take leave of the master.

Conclusion

Poincaré lived during that very specific moment in the history of science when randomness, in a more and more insistent way, challenged the beautiful deterministic edifice of Newton's and Laplace's cosmology which had dominated scientific thinking for centuries. A conference by Paul Langevin in 1913 [48] shows the extent of this challenge, paralleling the introduction of probabilities and a drastic change in our comprehension of the structural laws of matter. Such a penetrating mind as Poincaré could not have lived this irruption otherwise than as a traumatic one, that he had to face with the means he had at his disposal. These means, as we saw, had not yet reached the degree of power necessary to deal with many problems raised by modern physics. Let us recall one of the master's apothegms:

> 'Physics gives us not only an occasion for problem solving: it helps us find the means of solution, and in two ways. It points us in the direction of the solution, and suggests how to reason our way there.'[22]

And the new physical theories with which Poincaré was confronted suggested developing the theory of probability in the first place – a suggestion which can be also found, but in a slightly different perspective, among the problems Hilbert expounded during the Paris Congress of 1900. Therein lies the apparent paradox which puzzled the mathematician at the turn of the century: the hesitation and reluctance in the face of problems raised by statistical mechanics, the somewhat uncertain attempts to give solid bases to the theory of probability, the seemingly limited taste for new mathematical techniques, in particular measure theory and Lebesgue's integration, though they could have provided decisive tools to tackle numerous problems. Poincaré, as we said, remained a man of the 19th century, maybe in the same way as Klein had mischievously presented Gauss as a scientist of the 18th century. Naturally, in Gauss' case, the irony came from the fact that he had lived two thirds of his life in the 19th century, whereas death surprised Poincaré at the beginning of the 20th century. But we may speculate – although *not* here! – on the manner in which our hero would have adapted to transformations in the scientific world picture. We have seen that, following the example of his glorious predecessor, Poincaré sowed widely, and the spectacular blossoms of many of his ideas inspired countless researchers after his death. As for probabilities, I think one can sum up the measure of his influence as follows: he began to extract the domain from the grey zone to which it had been confined by almost all French mathematicians, he initiated methods that flourished when they integrated more powerful mathematical theories, he convinced Borel of the importance of certain questions, to the study of which he eventually devoted an enormous amount of energy. For a rather marginal subject in Poincaré's works, such a contribution appears far from negligible.

---

[22]'La physique ne nous donne pas seulement l'occasion de résoudre des problèmes; elle nous aide à en trouver les moyens, et cela de deux manières. Elle nous fait pressentir la solution; elle nous suggère des raisonnements.' ([73], p. 152)

References

[1] Appell, Paul, Darboux, Gaston, Poincaré, Henri: Examen critique des divers systèmes ou études graphologiques auxquels a donné lieu le bordereau. Rapport à la Cour de Cassation. Electronic Journal for History of Probability and Statistics (www.jehps.net) **1**, 1 (2005).

[2] Atten, Michel: La nomination de H. Poincaré à la chaire de physique mathématique et calcul des probabilités de la Sorbonne. Cahiers du séminaire d'histoire des mathématiques **9**, 221–230 (1988).

[3] Bachelier, Louis: Théorie de la spéculation. Annales Scientifiques de l'École Normale Supérieure, **17**, 21–86 (1900).

[4] Barberousse, Anouk: *La mécanique statistique de Clausius à Gibbs.* Paris, Belin, 2002.

[5] Barberousse, Anouk: La valeur de la connaissance approchée. L'épistémologie de l'approximation d'Émile Borel. Revue d'Histoire des Mathématiques **14**, fascicule 1, 53–75 (2008).

[6] Barbut, Marc, Locker, Bernard, and Mazliak, Laurent: *Paul Lévy – Maurice Fréchet, 50 ans de correspondance.* Hermann, Paris, 2004.

[7] Barrow-Green, June: *Poincaré and the Three Body Problem, History of Mathematics.* Vol. 11, American Mathematical Society – London Mathematical Society, 1997.

[8] Beranek, Jan: *Bohuslav Hostinský* (1884–1951). Časopis pro pěstování matematiky **109**, 442–448 (1984).

[9] Bertrand, Joseph: *Calcul des Probabilités.* Gauthier-Villars, 1889 (available on http://gallica.bnf.fr).

[10] Billingsley, Patrick: *Probability and Measure* (3rd ed.). Wiley and Sons, 1995.

[11] Borel, Émile: Sur quelques points de la théorie des fonctions. Annales Scientifiques de l'École Normale Supérieure **3**, 12, 9–55 (1895).

[12] Borel, Émile: *Leçons sur la théorie des fonctions.* Gauthier-Villars, Paris, 1898.

[13] Borel, Émile: Remarques sur certaines questions de probabilités. Bulletin de la Société Mathématique de France **33**, 123–128 (1905).

[14] Borel, Émile: Sur les principes de la théorie cinétique des gaz. Annales Scientifiques de l'École Normale Supérieure **23**, 9–32 (1906).

[15] Borel, Émile: *Eléments de la théorie des probabilités.* Hermann, Paris, 1909.

[16] Borel, Émile: Sur le battage des cartes. Comptes-Rendus de l'Académie des Sciences de Paris **154**, 23–25 (1912).

[17] Borel, Émile: Mécanique statistique, d'après l'article allemand de P. Ehrenfest et T. Ehrenfest, Encyclopédie des Sciences Mathématiques, Tome IV, Vol. 1, 188–292, 1915 (réédition J. Gabay, 1991; on-line at gallica.bnf.fr).

[18] Borel, Émile: *Valeur pratique et philosophie des probabilités, Traité du calcul des probabilités et leurs applications.* (Émile Borel, editor), Gauthier-Villars, 1939.

[19] Borel, Émile: *Œuvres, Introduction et Bibliographie par M. Fréchet.* 4 Tomes, CNRS, Paris, 1972.

[20] Bru, Bernard: Souvenirs de Bologne. Journal Soc. Fra. Stat. **144**, 1–2 (2003).

[21] Bru, Bernard: Les leçons de calcul des probabilités de Joseph Bertrand. Electronic Journal for History of Probability and Statistics (www.jehps.net) **2**, 2 (2006).

[22] Bru, Bernard: Une introduction analytique à la Théorie analytique. Hermann Laurent (1873), Electronic Journal for History of Probability and Statistics (www.jehps.net) **8** (2012).

[23] Bru, Bernard: Les probabilités dénombrables à la portée de tous. To appear (2012).

[24] Cartan, Elie: Les groupes bilinéaires et les systèmes de nombres complexes, Annales de la Faculté des Sciences de Toulouse, XII, B1-B99 (1898).

[25] Cartier, Pierre: *Poincaré's Calculus of probability* in [28].

[26] Carvallo, Emmanuel: *Le calcul des probabilités et ses applications.* Gauthier-Villars, Paris, 1912.

[27] Cavaillès, Jean: Du Collectif au Pari. Revue de Métaphysique et de Morale **XLVII**, 139–163 (1940).

[28] Charpentier, Eric, Ghys, Etienne, and Lesne, Annick (Eds.): *The scientific legacy of Poincaré, History of Mathematics.* Vol. 36, American Mathematical Society – London Mathematical Society, 2010.

[29] Courtault, Jean-Michel, and Kabanov, Yuri: *Louis Bachelier. Aux origines de la finance mathématique.* Presses Univ. Franc-Comtoises, 2002.

[30] Crawford, Elisabeth, and Olff-Nathan, Josiane (Eds.): *La science sous influence: l'Université de Strasbourg, enjeu des conflits franco-allemands.* Strasbourg, La Nuée bleue, 2005.

[31] Décaillot, Anne-Marie: *Cantor et la France.* Kimé, 2008.

[32] Durand, Antonin, and Mazliak, Laurent: *Revisiting the sources of Borel's interest for probability, Continued Fractions, Social involvement, Volterra's Prolusione. Centaurus*, 2011.

[33] Ehrenfest, Tanya and Paul: *Mécanique Statistique.* Traduit et complété par É. Borel à partir de la version allemande. Encyclopédie des mathématiques pures et appliquées. Jules Molk, éd., Tome IV, Vol. 1, 188–292, 1915.

[34] Fréchet, Maurice: Remarque sur les probabilités continues. Bulletin des sciences mathématiques **45**, 87–88 (1921).

[35] Fréchet, Maurice: *Recherches théoriques modernes sur le calcul des probabilités.* Livre 2, Méthode des fonctions arbitraires. Théorie des évènements en chaine dans le cas d'un nombre fini d'états possibles, *Traité du calcul des probabilités et de ses applications*, Gauthier-Villars (1938)

[36] Gargani, Julien: *Poincaré, le Hasard et l'étude des systèmes complexes.* L'Harmattan, 2012

[37] Gibbs, Josiah W.: *Elementary Principles in Statistical Mechanics.* Scribner, 1902.

[38] Hadamard, Jacques: Note de lecture sur J. Gibbs, "Elementary Principles in Statistical Mechanics". Bull. Amer. Math. Soc. **12**, 194–210 (1906); également: Bulletin des sciences mathématiques **30**, 161–179 (1906).

[39] Hadamard, Jacques: Sur le battage des cartes. Comptes-Rendus de l'Académie des Sciences de Paris **185**, 5–9 (1927).

[40] Havlova, Veronika, Mazliak, Laurent, and Šišma, Pavel: Le début des relations mathématiques franco-tchécoslovaques vu à travers la correspondance Fréchet-Hostinský. Electronic Journal for History of Probability and Statistics (www.jehps.net) **1**, 1 (2005).

[41] Hawkins, Thomas: *Lebesgue's theory of integration.* Chelsea AMS, 1975.

[42] Hostinský, Bohuslav: Sur les probabilités relatives aux transformations répétées. Comptes-Rendus de l'Académie des Sciences de Paris **186**, 59–61 (1928).

[43] Hostinský, Bohuslav: Nové řešení Buffonovy úlohy o jehle (New solution of Buffon problem on needle), *Rozpravy Ceské Akademie*, XXVI, II, 13, 1917 (French translation: Sur une nouvelle solution du problème de l'aiguille, Bulletin des sciences mathématiques **44**, 126–136 (1920)).

[44] Hostinský, Bohuslav: *O činnosti Karla Vorovky ve filosofii matematiky.* Ruch filosofický **8**, 65–71 (1929).

[45] Kamlah, Andreas: Probability as a quasi-theoretical concept – J.V. Kries' sophisticated account after a century. Erkenntnis **19**, 239–251 (1983).

[46] Kamlah, Andreas: *The Decline of the Laplacian Theory of Probability, in The Probabilistic Revolution* (Volume 1). Edited by L. Krüger, L.J. Daston and M. Heidelberger, Massachusetts Institute of Technology, pp. 91–116, 1987.

[47] Kelvin, Lord (J.J. Thomson): On a Decisive Test-case disproving the Maxwell–Boltzmann Doctrine regarding Distribution of Kinetic Energy. Philosophical Transactions of the Royal Society **51**, 397–399 (1892).

[48] Langevin, Paul: La Physique du discontinu. Conférence à la Société française de Physique le 27 novembre 1913. Republished in: Langevin, Paul: *La Physique depuis Vingt ans*, Doin, 189–264 (1923).

[49] Lebesgue, Henri: Sur une généralisation de l'intégrale définie. Comptes-Rendus de l'Académie des Sciences de Paris **132**, 1025–1028 (1901).

[50] Lévy, Paul: Sur le rôle de la loi de Gauss dans la théorie des erreurs. Comptes-Rendus de l'Académie des Sciences de Paris **174**, 855–857 (1922).

[51] Lindeberg, Jarl Waldemar: Sur la loi de Gauss. Comptes-Rendus de l'Académie des Sciences de Paris **174**, 1400–1402 (1922).

[52] Mansuy, Roger, and Mazliak, Laurent: Introduction au rapport de Poincaré pour le procès en cassation de Dreyfus en 1904. Electronic Journal for History of Probability and Statistics (www.jehps.net) **1**, 1 (2005).

[53] Mansuy, Roger, and Mazliak, Laurent: *L'analyse graphologique controversée d'Alphonse Bertillon dans l'affaire Dreyfus. Polémiques et réflexions autour de la figure de l'expert.* In Pierre Piazza: Alphonse Bertillon, aux origines de la police scientifique, ed. Karthala, 2011.

[54] Марков, АндрейА.: К вопоросу о разорении игроков (A.A.Markov: On the question of the gamblers' ruin). Bulletin de la Société Mathématique de Kazan, Série 2, Tome XIII, 38–45, 1905.

[55] Mazliak, Laurent: On the exchanges between Hostinský and Doeblin. Revue d'Histoire des Maths **13**, 155–180 (2007) *and* Electronic Journal for History of Probability and Statistics (www.jehps.net) **3**, 1 (2007).

[56] Mazliak, Laurent: An introduction to Karel Vorovka's philosophy of randomness. Electronic Journal for History of Probability and Statistics (www.jehps.net) **3**, 2 (2007).

[57] Mazliak, Laurent: The Ghosts of the École Normale. Life, death and destiny of René Gateaux. To appear.

[58] Mazliak, Laurent: A study of a trajectory: Popoff, wars and ballistics. Almagest III, 1, May 2012.

[59] Mazliak, Laurent, and Sage, Marc: Au delà des réels. Borel et l'approche probabiliste de la réalité. To appear in Revue d'Histoire des Sciences (2013).

[60] Pavlincova, Helena: *Karel Vorovka.* Cesta matematika k filosofii, Filosofia, Praha, 2010.

[61] Pier, Jean-Paul: Henri Poincaré croyait-il au calcul des probabilités? Philosophia Scientiae **1**, 4, 69–83 (1996).

[62] Poincaré, Henri: Sur le problème des trois corps et les équations de la mécanique. Acta Mathematica **13**, 1–270 (1890).

[63] Poincaré, Henri: Le problème des trois corps, Revue générale des sciences pures et appliquées **2**, 1, 1–4 (1891).

[64] Poincaré, Henri: *Thermodynamique.* Georges Carré, Paris, 1892.

[65] Poincaré, Henri: *Méthodes nouvelles de la Mécanique céleste, I. Solutions périodiques; non-existence des intégrales uniformes; solutions asymptotiques.* (1892); *II. Méthodes de MM. Newcomb, Gyldén, Lindstedt et Bohlin.* (1893); *III. Invariants intégraux; solutions périodiques du deuxième genre; solutions doublement asymptotiques.* Gauthier-Villars, Paris, 1892–99, 1899.

[66] Poincaré, Henri: Le Mécanisme et l'expérience. Revue de Métaphysique et de Morale **1**, 534–537 (1893).

[67] Poincaré, Henri: Sur la théorie cinétique des gaz. Revue générale des sciences pures et appliquées **5**, 11, 513–521 (1894).

[68] Poincaré, Henri: *Théorie analytique de la propagation de la chaleur*, Georges Carré, Paris, 1895.

[69] Poincaré, Henri: *Le calcul des Probabilités.* Georges Carré, Paris, 1896.

[70] Poincaré, Henri: Réflexions sur le calcul des probabilités. Revue générale des sciences pures et appliquées **10**, 262–269 (1899).

[71] Poincaré, Henri: *La science et l'hypothèse.* Flammarion, 1902.

[72] Poincaré, Henri: Sur l'intégration algébrique des équations linéaires et les périodes des intégrales abéliennes, Journal de Mathématiques pures et appliquées, **5**, 9, 139–212 (1903).

[73] Poincaré, Henri: *La valeur de la Science.* Flammarion, 1905.

[74] Poincaré, Henri: Réflexions sur la théorie cinétique des gaz. Journal de Physique, 4ème série **V**, 369–403 (1906).

[75] Poincaré, Henri: Le hasard. Revue du mois **3**, 257–276 (1907).

[76] Poincaré, Henri: *Science et Méthode.* Flammarion, 1908.

[77] Poincaré, Henri: *Le calcul des Probabilités.* 2ème édition, Gauthier-Villars, Paris, 1912.

[78] Poincaré, Henri: Analyse des travaux scientifiques de Henri Poincaré, faite par lui même. Acta Mathematica **38**, 1–135 (1921).

[79] Poisson, Denis Syméon: *Recherches sur la probabilité des jugements en matière criminelle et en matière civile.* Bachelier, 1837.

[80] Попов, Кирил: Автобиография. Университетско Издателство "Св. Климент Охридски", София, 1993.

[81] Rollet, Laurent: Autour de l'Affaire Dreyfus: Henri Poincaré et l'Action politique. Revue Historique **CCXCVIII/3**, 49–101 (1999).

[82] Schneider, Ivo: *Laplace and Thereafter: The Status of Probability Calculus in the Nineteenth Century.* 191–214, 1987. In The Probabilistic Revolution (Volume 1), edited by L. Krüger, L.J. Daston and M. Heidelberger, Massachusetts Institute of Technology, p. 91–116, 1987.

[83] Seneta, Eugene: *Non-negative Matrices and Markov Chains.* 2nd Edition, Springer Series in Statistics, Springer-Verlag, New-York, 1981.

[84] Sheynin, Oscar B.: H. Poincaré's work on probability. Archives for History of Exact Sciences **42**, 131–171 (1991).

[85] Siegmund-Schultze, Reinhardt: *Maurice Fréchet à Strasbourg.* Chapter in [30].

[86] Stigler, Stephen M.: Laplace's 1774 Memoir on Inverse probability, Statistical Science, 1986, 1, 3, 359–378)

[87] Tait, Peter G.: Poincaré's Thermodynamics. Nature **45**, 245–246 (1892).

[88] Van der Vaart, Aad W.: Asymptotic Statistics. Cambridge University Press, 1998

[89] Von Kries, Johannes: *Die Principien der Wahrscheinlichkeitsrechnung, eine logische Untersuchung.* Akademische Verlagsbuchhandlung Mohr, 1886.

[90] Von Plato, Jan: Boltzmann's Ergodic Hypothesis. Archives for History of Exact Sciences **42**, 71–89 (1991).

[91] Von Plato, Jan: *Creating modern probability.* Cambridge University Press, 1994.

[92] Vorovka, Karel: Filosofický dosah počtu pravděpodobnosti. Česká mysl. **14**, 17–30 (1912).

[93] Vorovka, Karel: Poznámka k problému ruinováníhráčů (*A note to the problem of gamblers ruin*). Časopis pro pěstování matematiky, XLI, 1912.

[94] Vorovka, Karel: O pravděpodobnosti příčin (*On the probability of causes*), Časopis pro pěstování matematiky, XLIII, 1914.

[95] Wiman, Anders: Über eine Wahrscheinlichkeitsaufgabe bei Kettenbruchentwickelungen. Stockh. Öfv. **57**, 829–841 (1900).

Laurent Mazliak
Laboratoire de Probabilités et Modèles Aléatoires
Université Pierre et Marie Curie
4, Place Jussieu
F-75252 Paris Cedex 05
e-mail: `laurent.mazliak@upmc.fr`

Henri Poincaré, 1912–2012, 193–230

Poincaré Seminar 2012

Henri Poincaré and the Uniformization of Riemann Surfaces

François Béguin

Abstract. Uniformization of Riemann surfaces is one of the mathematical problems that have accompanied H. Poincaré all along his mathematical life. I shall evoke six "uniformization theorems" discovered by Poincaré. He proved the first of these six theorems when he was a twenty-six years old assistant professor in Caen, and the last one twenty-six years later, when he was one of the most celebrated scientists in the world.

The present text is in large part extracted from the book that Henri-Paul de Saint-Gervais has written about the history of the uniformization of Riemann surfaces ([St-Gervais2010]).

1. Introduction

Almost every modern textbook on Riemann surfaces contains the following statement of the "Uniformization Theorem":

Every simply connected Riemann surface is biholomorphically equivalent to the Riemann sphere, the complex plane, or the unit disc.

If one wishes to understand the history of the Uniformization Theorem in the XIXth century (from the work of N. Abel and C. Jacobi on elliptic curves in the 1830s till the proof of the Uniformization Theorem by Poincaré and P. Koebe in 1907), the best thing to do is certainly to forget this modern statement for a while.

The modern statement of the Uniformization Theorem suggest that, in order to understand all Riemann surfaces, it is enough to describe those that are simply connected. Indeed, given any Riemann surface S, one may apply the above Uniformization Theorem to the universal cover of S (which is by definition simply connected), and deduce that S is biholomorphically equivalent to the Riemann sphere, to a quotient of the complex plane by a lattice of translations, or to the quotient of the unit disc by a discrete group of hyperbolic isometries (i.e., by a Fuchsian group). But, when mathematicians started to wonder about the uniformization of Riemann surfaces in the middle of the XIXth century, the notion of universal cover had not arisen yet, and hyperbolic geometry did not exist[1]. The discovery of the concept of universal cover as well as the discovery of Fuchsian groups are two revolutions in the history of mathematics before which it was even impossible to dream of a uniformization theorem for *all* Riemann surfaces.

The above modern statement of the Uniformization Theorem presents it as a *classification result* (for Riemann surfaces, up to biholomorphism). However, in the XIXth century uniformization was certainly not a matter of classification – it was a matter of *parametrization*. A (real or complex) curve can be defined in two different ways: in an implicit way, by means of an equation $F(x, y) = 0$, or in a explicit way, by means of a parametrization $t \mapsto (x(t), y(t))$. In the XIXth century, uniformizing a Riemann surface consisted in finding a parametrization for a Riemann surface (which was typically given by an implicit algebraic equation). Under this viewpoint, the modern statement of the Uniformization Theorem tells us that any Riemann surface can be parametrized by the sphere, the complex plane, or the unit disc. Moreover, the parametrization is very "pleasant": this is a local biholomorphism and a covering map (but of course, it cannot be one-to-one, except when the Riemann surface under consideration is simply connected).

Last, the modern statement of the Uniformization Theorem concerns some geometrical objects: Riemann surfaces. In the XIXth century (in particular when uniformization was concerned), Riemann surfaces were thought as tools, rather than as abstract geometrical objects interesting in their own right. The words of D. Hilbert in his address to the International Congress of Mathematicians in 1900 are particularly significant in this respect. Hilbert speaks of a "relation between two complex variables", rather than of the Riemann surface defined by this relation:

> *As Poincaré was the first to prove, it is always possible to reduce any algebraic relation between two variables to uniformity by the use of automorphic functions of one variable. That is, if any algebraic equation in two variables be given, there can always be found for these variables two such single-valued automorphic functions of a single variable that their substitution renders the given algebraic equation an identity. The generalization of this fundamental theorem to any analytic non-algebraic relations whatever between two variables has likewise been attempted with success by Poincaré,* [...][2]

In any case, when Bernhard Riemann introduced the surfaces that today are named after him, his goal was certainly not to study some abstract geometrical objects[3];

---

[1]Actually, it did exist as an abstract construction. But nobody would have hoped hyperbolic geometry to play any kind of role in any "concrete" mathematical problem.

[2]To be honest, we must mention that Poincaré often uses a more geometrical language in this papers. For example, in some of his papers, he states the Uniformization Theorem in the following terms: *"The coordinates of the points of an arbitrary algebraic curve can be expressed as Fuchsian functions of an auxiliary variable."*

[3]Although he never hesitates, when necessary, to treat Riemann surfaces as topological surfaces, which he can cut and paste along closed curves.

the Riemann surfaces he considers are defined by an algebraic equations, more generally, are associated to mutli-valued functions. The *uniformization problem* did not came to light as mathematicians were studying Riemann surfaces; it arose as soon as mathematicians discovered the existence of *muti-valued functions*[4]. One might even say that Riemann surfaces were defined in order to understand better the uniformization problem! Here is the statement of the "Uniformization Theorem" which was proved by Poincaré in 1883:

> *Let y be an arbitrary multi-valued analytic function of a variable x. One can always find a variable z such that x and y are single-valued functions of z.*

Of course, this statement can be rephrased in purely geometrical terms, by introducing the Riemann surface associated to the germ of the analytic function y at some (arbitrary) point. Expressing x and y as single-valued functions of an auxiliary variable z amounts to finding a (single-valued) parametrization of this Riemann surface. This is what Poincaré actually does in order to prove the above statement. It remains that the goal of Poincaré was to establish a result on multivalued functions (Poincaré's paper is called "*On a theorem of the general theory of functions*"), and that Riemann surfaces only appear as tools in his paper.

The path which led Poincaré to the uniformization of algebraic Riemann surfaces in 1881–82 is also significative. In the beginning of his career, Poincaré was interested in differential equations. In 1881, a paper of L. Fuchs caught his attention, and focussed his interest on linear differential equations of the second order with meromorphic coefficients. The solutions of such differential equations are naturally multi-valued (the actual value of the solution changes when the variables follows a path that goes around a pole of the coefficients of the equation). Poincaré discovered (and he was elated by this discovery) that the solutions of the equation can nevertheless be expressed as *single-valued* functions of an auxiliary variable. This allowed him to prove that every algebraic Riemann surface (of genus at least two) can be uniformized by the unit disc... but this incredible result was not his initial goal[5].

Poincaré is the main character of the history of the Uniformization Theorem. In 1880, he suddenly realizes that hyperbolic geometry allows one to construct meromorphic functions with incredibly rich period lattices, and understands that those new functions can play for algebraic curves of arbitrary degree the same role as elliptic function played for non-singular cubic curves. After this spark of genius, a uniformization theorem for *all* algebraic Riemann surfaces becomes possible.

[4]Nowadays, the phrase "multi-valued function" sounds as an oxymoron: a "function" in the modern sense can not be "multi-valued", since it assigns a single value to each point in its domain. The association of these two words is nevertheless meaningful, and underlines the problem posed by the mere existence of such objects.

[5]Even afterwards, Poincaré was much more excited by his result on linear differential equations, than by the consequences in terms of algebraic Riemann surfaces. To his friend Lecornu, he does not tell a word about algebraic Riemann surfaces, but enthusiastically declares: "I can solve all differential equations!".

This is not Poincaré, but rather Felix Klein, who will be the first to dare to state the uniformization theorem for general algebraic Riemann surfaces. Nevertheless, Poincaré was certainly the one who made such a result possible. Before Poincaré's work, nobody would have dreamed of such a general result.

At the end of 1882, Klein and Poincaré both consider that they have managed (independently from each other, and using quite different arguments) to set up convincing proofs of the Uniformization Theorem for algebraic Riemann surfaces. Probably exhausted by the competition with Poincaré, Klein will shortly fall ill and face a two-years nervous breakdown. This will – in his own words – be the end of his productive period[6]. On the contrary, Poincaré is galvanized by his accomplishment, and tries to do something that certainly seemed foolish at that time: he tries to uniformize *all* Riemann surfaces (algebraic or not). Only one year later, he will obtain a weaker, but completely general Uniformization Theorem. As stated by Hilbert, Poincaré result amounts to "uniformizing an arbitrary algebraic relation between to complex variables".

Twenty-five years later, it was once again Poincaré (simultaneously with Paul Koebe) who finally proved the general Uniformization Theorem that we know today. It is true that Poincaré's proof is not as rigorous as Koebe's, but the physical ideas which guide that proof are illuminating.

The present text has a very limited ambition. I certainly do not intend to present the history of the Uniformization Theorem. Neither do I pretend to analyze the role played by Poincaré in the conquest of this fundamental result. I just consider this text as a modest contribution to the commemoration of 100 years since Poincaré's death by sketching some of his fundamental contributions to the uniformization problem. More precisely, I will evoke six "uniformization theorems", all proved by Poincaré at several stages of his exceptional career: first the uniformization of the Riemann sphere minus a finite number of real points, then the uniformization of algebraic Riemann surfaces modulo a finite number of points, then the "true" uniformization of algebraic Riemann surfaces by means of the "continuity method", then the uniformization of (multi-valued) analytic functions, once again the uniformization of Riemann surfaces (but this time, by solving the Liouville equation), and finally *the* Uniformization Theorem that we know nowadays. I hope that these six statements, together with a few hints of their proofs, will allow the reader to perceive the astounding creative power of Poincaré, and the incredible diversity of the facets of his genius.

Almost all the material of the following pages was extracted from a book, written together with fourteen colleagues, on the occasion of the celebration of the 100th birthday of the Poincaré–Koebe Uniformization Theorem ([St-Gervais2010]). I would like to express my gratitude to my co-authors who allowed me to shamelessly borrow what I needed here.

[6]He was only 33 years old!

2. Uniformization modulo a finite number of points

2.1. The apparition of hyperbolic geometry

At the end of his life, Poincaré told the story of his discovery of *Fuchsian functions* ([Poincaré1908]).

In 1880, a paper of Fuchs ([Fuchs1866]) caught Poincaré's attention. This paper dealt with linear second-order differential equations with meromorphic coefficients. The solutions of such an equation are multi-valued functions (the actual value of a solution changes when the variable makes one turn around a pole of one of the coefficients of the equation). In order to understand the behaviour of a linear basis of solutions (v_1, v_2), it is enough to study the ratio $w = v_1/v_2$ (which is of course a multi-valued function as well). Guided by the analogy with elliptic functions, one tries to consider the "inverse" of the multi-valued function w : if w were one-to-one, its inverse w^{-1} would be a "true function" (i.e., a single-valued function), and the multi-valuedness of w would translate as a periodicity of w^{-1}. Alas, Poincaré soon realized that the function w is always "very widely multi-valued", and thus, that its inverse w^{-1} (assuming that it exists) would have to admit an incredibly complicated "lattice of periods".

Poincaré was first convinced that meromorphic functions with such a complicated lattices of periods could not exist. But, a few days later, he unexpectedly managed to construct examples of such functions, which he would call *Fuchsian functions.* He was soon distracted from his mathematical preoccupations by a geological trip organized by the *École des Mines.* It was during this trip that "*the idea came to [him], without anything in [his] former thoughts seeming to have paved the way for it, that the transformations [he] had used to define the Fuchsian functions were identical to those of non-Euclidean geometry*". In other words, Poincaré suddenly realized that the "Fuchsian functions" he had constructed were meromorphic functions defined on the unit disc and invariant under a discrete group of transformations which are isometries for the hyperbolic metric (nowadays called *Poincaré metric*) on the disc.

This "illumination", which revealed to Poincaré the existence of a link between the uniformization of certain multi-valued functions (and, therefore, the uniformization of certain Riemann surfaces) and hyperbolic geometry, opened a road towards to the Uniformization Theorem. Of course, at that stage, Poincaré could only deal with very particular multi-valued functions (more precisely, quotients of solutions of very special linear differential equations). But Poincaré was now in the possession of a geometrical tool which allowed him to construct "new transcendental functions", with astonishing new properties, that were natural candidates to be uniformizing functions.

From that moment on, Poincaré had a work plan in three points:

1. construct discrete groups of hyperbolic isometries of the unit disc (which Poincaré would soon call *Fuchsian groups*; try to understand the moduli space of all such groups;

2. for every Fuchsian group Γ, build Fuchsian functions associated to Γ, i.e., meromorphic functions on the unit disc that are invariant under all the elements of Γ; try to understand the space of all such functions;
3. finally, show that Fuchsian functions constructed in this manner allow one to solve many (all?) linear differential equations, and thus to uniformize many (all?) Riemann surfaces[7].

2.2. Construction of Fuchsian functions

Recall that the Riemann sphere is the Riemann surface (homeomorphic to the standard two-dimensional Euclidean sphere) obtained by "adding a point at infinity to the complex plane"; we shall denote it by $\widehat{\mathbb{C}}$. Also recall that the biholomophic self-maps of the Riemann sphere $\widehat{\mathbb{C}}$ are exactly the homographies with complex coefficients, i.e., the transformations of the type $z \mapsto (az+b)/(cz+d)$ with $a, b, c, d \in \mathbb{C}$ and $ad-bc \neq 0$. The group of all such biholomorphic self-maps is isomorphic to $\mathrm{PSL}(2,\mathbb{C})$. It turns out that the subgroup made of the maps which leaves the unit disc $\mathbb{D}$ invariant coincides with the group of isometries of the standard (Poincaré) hyperbolic metric on $\mathbb{D}$. This group is isomorphic to $\mathrm{PSL}(2,\mathbb{R})$.

Definition 1. A *Fuchsian group* is a discrete subgroup of the group of hyperbolic isometries of the unit disc $\mathbb{D}$.

Let us leave for a moment the construction of Fuchsian groups, in order to explain how Poincaré builds Fuchsian functions associated to such a group. Let Γ be a Fuchsian group.

Definition 2. A Fuchsian function associated with a group Γ is a meromorphic function $g : \mathbb{D} \to \widehat{\mathbb{C}}$ which is invariant under Γ, i.e., which satisfies $g \circ \varphi(z) = g(z)$ for every $\varphi \in \Gamma$.

Of course, given any meromorphic function g on the disc $\mathbb{D}$, one may naively try to build a Γ-invariant function by considering the series

$$\sum_{\varphi \in \Gamma} g \circ \varphi(z).$$

Unfortunately, this series does not converge (except in the trivial situations). To overcome this problem, Poincaré follows the analogy with the theory of elliptic functions. He first seeks for functions that are not exactly Γ-invariant, but rather "Γ-quasi-invariant".

Definition 3. Let ν be a positive real number. A *meromorphic automorphic form of weight* ν associated with the Fuchsian group Γ is a meromorphic function g on the unit disc $\mathbb{D}$ which satisfies

$$g \circ \varphi(z) \cdot (\varphi'(z))^{\nu} = g(z)$$

for all $\varphi \in \Gamma$.

[7]Although differential equations were Poincaré's main motivation, here we will leave them aside, in order to focus on the uniformization aspects.

Poincaré proves the following result (see [Poincaré1882a] or [St-Gervais2010, Chapter VI]):

Theorem 4 (Poincaré, 1881). *Let g be a meromorphic function on the Riemann sphere $\widehat{\mathbb{C}}$ which does not have any pole on the boundary of the unit disc $\mathbb{D}$. For every $\nu \geq 2$, the séries*

$$\theta(z) := \sum_{\varphi \in \Gamma} f \circ \varphi(z) \cdot (\varphi'(z))^{\nu} \tag{1}$$

converges uniformly on any compact subset of $\mathbb{D}$. The sum of this series is a meromorphic automorphic form of weight ν associated to the group Γ.

It only remains to notice that the quotient $g = \theta_1/\theta_2$ of two meromorphic automorphic forms θ_1, θ_2 with the same weight is a Γ-invariant meromorphic function, i.e., a Fuchsian function associated to the Fuchsian group Γ. If the automorphic forms θ_1, θ_2 have been chosen in such a way that they do not have exactly the same poles, then the Fuchsian function g is not constant. This shows that *one can construct non-constant Fuchsian function associated to the Fuchsian group* Γ.

Poincaré goes much further in [Poincaré1882a, §4]: he describes the space of all the Fuchsian functions associated to the Fuchsian group Γ. In the sequel, Γ is assumed to be finitely generated, and such that the hyperbolic area of the surface $\Gamma \setminus \mathbb{D}$ is finite (this is equivalent to requiring that the unit circle is contained in the closure of the orbit $\Gamma \cdot x$ of any point $x \in \mathbb{D}$). Poincaré observes that any two Fuchsian functions associated to the group Γ are related by an algebraic equation. Indeed, let $z \mapsto x(z)$ and $z \mapsto y(z)$ be two Fuchsian function associated to Γ. The inverse of x is of course multi-valued. But, the hypotheses made on Γ prevent x^{-1} from being "too widely multi-valued": it has only finitely many local determinations (branches). If one denotes by $z_1(x), \ldots, z_d(x)$ these local determinations, then any symmetric polynomial of $y(z_1(x)), \ldots, y(z_i(x))$ is a "true" (i.e., single-valued) meromorphic function on the Riemann sphere, i.e., a rational function of x. Therefore, the equation

$$y^d - \sigma_1(x)y^{d-1} + \cdots + (-1)^d \sigma_d(x) = 0$$

(where $\sigma_1, \ldots, \sigma_d$ are the elementary symmetric polynomials of $y(z_1(x)), \ldots, y(z_i(x)))$ is an algebraic equation relating x and y. Using the primitive element theorem, Poincaré deduces ([Poincaré1884]) that

> *All these functions* [the Fuchsian functions associated to the Fuchsian group Γ] *can be expressed as rational functions of two of them, which I will denote by x and y. Moreover, these two functions x and y will satisfy an algebraic relation $\varphi(x, y) = 0$.*

Then Poincaré deduces that the quotient $\Gamma \setminus \mathbb{D}$ can be identified[8] to the algebraic curve $\varphi(x, y) = 0$, via the map $z \mapsto (x(z), y(z))$.

2.3. Uniformization of the Riemann sphere minus a finite number of real points

In a short paper dated 18th May 1881 ([Poincaré1881c]), Poincaré announces that he is able to prove the following result[9]:

Theorem 5 (First Uniformization Theorem: uniformization of the Riemann sphere minus a finite number of real points. Poincaré, 1881). *Given any finite set of points $p_1, \dots, p_n$ on the real line, the Riemann sphere $\widehat{\mathbb{C}}$ minus the points $p_1, \dots, p_n$ can be uniformized by the unit disc*[10].

Poincaré sketched the proof of this result in a note he sent to the French Academy of Sciences three weeks later ([Poincaré1881d]). I will detail this proof in the case where the integer n in equal to 4; the general case follows from roughly the same arguments.

So let us assume that $n = 4$. We have to prove that the Riemann sphere minus four arbitrary real points $p_0, \dots, p_3$ can be uniformized by the unit disc. Using a well-chosen automorphism of $\widehat{\mathbb{C}}$, we can send three of the four points $p_0, \dots, p_3$ to 0, 1 and ∞, and place the fourth point in the real interval $]0, 1[$. So we are left to prove that, for every z in the interval $]0, 1[$, there exists a Fuchsian group Γ such that the quotient $\Gamma \setminus \mathbb{D}$ is biholomorphic to $\widehat{\mathbb{C}} - \{0, z, 1, \infty\}$.

In order to prove this, we have to construct Fuchsian groups. Recall that the geodesics of the standard hyperbolic metric of the disc are the arcs of circles which are orthogonal to the boundary of the disc. Let us consider six pairwise distinct points $s_1, \dots, s_6$ on the unit circle (we assume that the indexation $s_1, \dots, s_6$ is done in the clockwise direction on the circle). Let H be the hyperbolic hexagon with vertices $s_1, s_2, s_3, s_4, s_5, s_6$ (the vertices of this hexagon are on the unit circle, i.e., "at infinity", and therefore do not belong to the hexagon itself). There exist a unique isometry ϕ_2 which maps the geodesic $]s_2, s_3[$ into the geodesic $]s_2, s_1[$, a unique isometry ϕ_4 which maps the geodesic $]s_4, s_5[$ into the geodesic $]s_4, s_3[$, and a unique isometry ϕ_6 which maps the geodesic $]s_6, s_1[$ into the geodesic $]s_6, s_5[$ (Figure 1). Let Γ be the group of isometries generated by ϕ_2, ϕ_4, ϕ_6. It is not hard to see that Γ is a Fuchsian group and that the quotient space $\Gamma \setminus \mathbb{D}$ is nothing but

[8] Poincaré seems to feel a bit uncomfortable with the precise meaning of this identification, and remains quite vague about it. Actually, the quotient $\Gamma \setminus \mathbb{D}$ is biholomorphic to the Riemann surface associated to the algebraic curve of equation $\varphi(x, y) = 0$, that is, to the curve obtained by desingularizing and compactifying $\{(x, y) \in \mathbb{C}^2 \mid \varphi(x, y) = 0\}$.

[9] Theorem 5 is a translation of Poincaré's statement in modern and geometrical terms. The direct translation of Poincaré's original statement would be:

> *By introducting zeta-Fuchsian functions corresponding to these functions $f(z)$, one can solve all the linear equations with rational coefficients, whose singular points lie on the real line.*

[10] In other words, $\widehat{\mathbb{C}} - \{p_1, \dots, p_n\}$ is biholomorphic to the quotient of the unit disc by a Fuchsian group.

the Riemann surface obtained by considering the hexagon H and gluing pairwise the opposite edges of H. This Riemann surface is obviously homeomorphic to the sphere minus four points (see Figure 1). H. Schwarz had already proved in 1870 that, up to biholomorphism, the Riemann sphere $\widehat{\mathbb{C}}$ is the only Riemann surface homeomorphic to a sphere. Since any triple of pairwise distinct points of the Riemann sphere can be mapped on $0, 1, \infty$ by a well-chosen biholomorphism of $\widehat{\mathbb{C}}$, one concludes that the quotient $\Gamma \setminus \mathbb{D}$ is biholomorphic to $\widehat{\mathbb{C}} - \{0, 1, \infty, z\}$ for a certain point z.

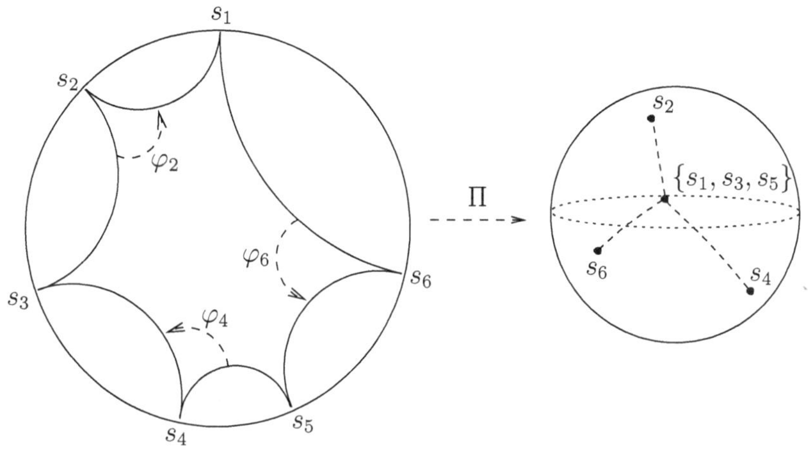

FIGURE 1. Uniformization of the sphere minus 4 points

Let us now focus on the particular case where the hexagon H is symmetric with respect to the axis $]s_2, s_5[$. Composing if necessary by a well-chosen isometry of $\mathbb{D}$, we may assume that $s_2 = 1$, $s_4 = i$, and $s_5 = -1$ (in particular, $]s_2, s_5[$ is nothing but the real axis). Then $s_3 = e^{i\theta}$ for a certain $\theta \in]0, \pi/2[$ and (since H is symmetric with respect to $]s_2, s_5[$, i.e., with respect to the real axis), $s_6 = \overline{s_4} = -i$ and $s_1 = \overline{s_3} = e^{-i\theta}$. Therefore, the hexagon H, the group Γ, and the point z depend only on the parameter $\theta \in]0, \pi/2[$; from now on, we shall denote them by $H_\theta, \Gamma_\theta, z_\theta$. The hexagon H_θ is invariant under the conjugation $z \mapsto \overline{z}$; therefore, the Riemann surface $\Gamma_\theta \setminus \mathbb{D} \simeq \widehat{\mathbb{C}} - \{0, 1, \infty, z_\theta\}$ admits an anti-holomorphic involution. It follows that the point z_θ must be on the real axis. We may assume that z_θ is in the real interval $]0, 1[$ (otherwise, we exchange the roles of the four points). It remains to prove that, when θ ranges over $]0, \pi/2[$, the point z_θ ranges over the whole interval $]0, 1[$.

One first checks that z_θ depends continuously on θ. Poincaré's proof of this fact consists in noticing that the isometries ϕ_2, ϕ_4, ϕ_6 obviously depend continuously on θ. Hence, for a given meromorphic function f, the automorphic form given by the series (1) will depend continuously on θ. Since the quotient

$\Gamma_\theta \setminus \mathbb{D} \simeq \widehat{\mathbb{C}} - \{0, 1, \infty, z_\theta\}$ can be identified to the algebraic curve defined by the relation between two Fuchsian functions, it follows that the point z_θ will depend continuously on θ.

Then one checks that the point z_θ tends toward the point 0 (resp. 1) when θ goes to 0 (resp. $\pi/2$). The proof uses the same arguments as those of the preceding paragraph. When θ goes to 0, the hexagon H_θ tends toward the 4-gon with the vertices $1, i, -1, -i$. Hence, for a given meromorphic function f, the automorphic form given by the series (1) will converge (uniformly on every compact set of $\mathbb{D}$) to the automorphic form associated given by a similar formula and associated to the group Γ_0 generated by the isometries which identifies pairwise the edges of the 4-gons $1, i, -1, -i$. It follows, that the Fuchsian function obtained by considering the quotient of two such automorphic forms will tend to the Fuchsian function defined in the same way, but associated to the group Γ_0. One deduces that z_θ will tend to 0 as θ will go to 0, and this concludes the proof of Theorem 5 (in the case $n = 4$).

The proof described above, in which Poincaré deforms continuously Fuchsian groups, is a prefiguration of what Klein and Poincaré will soon call the *continuity method.*

2.4. Uniformization of Riemann surfaces modulo a finite number of points

In his note to the Comptes Rendus of the French Academy of Sciences dated 8th August 1881 ([Poincaré1881e]), Poincaré writes without any hesitation:

> *One concludes*
> 1. *that any linear differential equation with algebraic coefficients can be solved by means of zeta-Fchsian functions.*
> 2. *that the coordinates of the points of an arbitrary algebraic curve can be expressed as Fuchsian functions of an auxiliary variable.*

In other words, he announces *that he can uniformize every algebraic curve*!

Actually, he is exaggerating. Here is the precise statement that Poincaré actually proves in his note:

Theorem 6 (Second Uniformization Theorem: uniformization modulo a finite number of points, Poincaré, 1881). *For every algebraic Riemann surface S, there exist a finite number of points $p_1, \dots, p_n \in S$ such that the Riemann surface $S - \{p_1, \dots, p_n\}$ is uniformized by the unit disc*[11].

Therefore, Poincaré can only uniformize algebraic curves *up to a finite number of points.* This is already an extremely impressive result : Poincaré can treat *all algebraic curves.* This is nevertheless nothing but an easy corollary of the uniformization of the sphere minus a finite number of real points (Theorem 5).

[11] In more modern terms, "such that $S - \{p_1, \dots, p_n\}$ is biholomophic to the quotient of the unit disc by a Fuchsian group.

Poincaré gives a complete and carefully detailed proof[12] of this corollary in his note.

Let us consider an algebraic Riemann surface S. Let $f : S \to \widehat{\mathbb{C}}$ be a meromorphic (hence rational) function defined on S (for example, if $\varphi(x, y) = 0$ is an equation of S, the coordinate x provides a rational function on S). Let $E \subset \widehat{\mathbb{C}}$ be the set of all critical values of f. If E happens to be contained in the real line, then Theorem 6 follows immediately from Theorem 5. Indeed, in this case, one can lift (using f) the uniformization of $\widehat{\mathbb{C}} - E$ provided by Theorem 5, and get a uniformization of $S - f^{-1}(E)$. If E is contained in the real line (which is the general case), Poincaré replaces f by $P \circ f$, where P is a suitable polynomial. The set of the critical values of $P \circ f$ is the union of the set $P(E)$ and the set of the critical value of P. Therefore, the proof of Theorem 6 reduces to the following lemma:

Lemma 7. *For every finite subset E of the Riemann sphere $\widehat{\mathbb{C}}$, there exists a polynomial P such that $P(E)$ is contained in the real line, and so is the set of the critical values of P.*

The proof of this lemma is an elementary exercise, which we leave to the reader (or which can be found in [Poincaré1881e]).

3. The continuity method and the uniformization of algebraic curves

On June 11th 1881, Klein reads the first papers of Poincaré on Fuchsian functions ([Poincaré1881a, Poincaré1881b, Poincaré1881c]). He is irritated by the ignorance of Poincaré, who has obviously not read Riemann's *Habilitationsschrift*, nor his own paper on the subject. But he is also certainly extremely impressed by the results already obtained by Poincaré. The next day, he writes a letter to Poincaré, seemingly to point out some papers that cannot be ignored by any mathematician pretending to work on algebraic curves. Actually, the true aim of this letter is certainly to get in touch with a young mathematician, who has just made a dramatic entrance into a mathematical field that was one of Klein's favorites. This is the first of the twenty-six letters that Poincaré and Klein will exchange within one year, in which they will talk about Fuchsian groups, Fuchsian functions, and uniformization of algebraic curves.

During the spring of 1881, Poincaré found some quite malleable methods to construct Fuchsian groups ([Poincaré1881a, Poincaré1881b, Poincaré1881c]). The main subject[13] of the first letters exchanged by Poincaré and Klein concerns the number of free parameters in the construction of a Fuchsian group Γ such that

[12] Which is quite unusual for him!

[13] Leaving apart a quarrel concerning the names chosen by Poincaré for the mathematical objects he has just invented (*Fuchsian groups*, *Fuchsian functions*). Klein considers that Fuchs does not deserve such an honor, and does not hesitate to put pressure on Poincaré to drop this

$\Gamma\backslash\mathbb{D}$ is a closed surface of genus g. In these letters, the two great mathematicians conclude that such a group Γ depends (up to conjugacy) on $6g-6$ real parameters. On the other hand, Riemann had proved that an algebraic curve of genus g depends (up to birational equivalence) of $3g-3$ complex parameters. Therefore, the dimension of the moduli space[14] of closed Riemann surfaces of genus g uniformized by the disc is equal to the dimension of the moduli space of algebraic curves of genus g. As soon as they have observed this equality, Klein and Poincaré are convinced that all algebraic curves with negative Euler characteristic are uniformized by the unit disc. Both of them will announce this result in 1882, leaving the proof for later:

Theorem 8 (Third Uniformization Theorem: uniformization of algebraic curves. Klein, Poincaré, 1882). *Every algebraic Riemann surface with negative Euler characteristic is uniformized by the unit disc*[15].

The arguments considered by Poincaré and Klein in order to prove this result are of quite different nature (for example, Poincaré still thinks in terms of differential equations, whereas Klein approaches the problem in more geometrical terms). Nevertheless, the proofs they try to work out follow the same global strategy, which they call the *continuity method.* This method consists in proving that, in the moduli space of all closed Riemann surfaces of genus g, the subset formed by the (class of) Riemann surfaces that can be uniformized by the unit disc is open and closed. This would be enough to prove Theorem 8, since it is easy to exhibit one closed Riemann surface of genus g which is uniformized by the unit disc, and since the connectedness of the moduli space of all closed Riemann surfaces of genus g is considered as obvious by both Poincaré and Klein.

The "proofs" of Theorem 8 which Poincaré and Klein will finally publish in 1883–84 are not very convincing[16]. At the end of his life, Klein will recognize that neither him, nor Poincaré had found a complete proof of Theorem 8 ([Klein1921a, vol. 3, pages 577–586]). One can nevertheless credit Poincaré for an "almost-complete" (and magnificent) proof of the fact that the set of surfaces uniformized by the disc is open. We shall present this proof below.

3.1. The set of uniformizable surfaces is open

Given a Riemann surface S and a local holomorphic coordinate x defined on an open set U of S, Poincaré considers linear differential equations of order two that can be written in the form

$$\frac{d^2v}{dx^2}+hv=0, \tag{2}$$

terminology. Klein explains that *Schwarzian functions* (or *Kleinian functions*!) would be a much more appropriate terms.

[14] Although, Poincaré and Klein seek for the dimension of this space, they certainly did not prove that this space was a manifold.

[15] *that is to say,* is biholomophic to the quotient of the unit disc by a Fuchsian group.

[16] at the very least, they did not convince H.-P. de Gervais, despite many efforts made to understand those proofs.

where h is a holomorphic function of x. The set of all the solutions of such a differential equation is obviously a two-dimensional linear space. Moreover, if (v_1, v_2) is a linear basis of that space, then v_1 and v_2 can be recovered if one just knows their ratio $w = v_1/v_2$. For this reason, we will be interested in the ratio of two solutions rather than in the solutions themselves. Poincaré considers only *normal* differential equations: these are differential equations as above, such that the ratio w of two solutions can be extended to a meromorphic (multi-valued) function defined on the whole surface S (see [St-Gervais2010, part VIII.3] for a more formal definition, and [St-Gervais2010, Part IX.1] for a characterization of these equations due to Fuchs).

Let S be a Riemann surface and (E) be a normal equation on S. Let v_1, v_2 be linearly independent solutions of (E) and $w = v_1/v_2$ be their ratio. Of course, v_1, v_2, w are not "true" functions: they are multi-valued. These multi-valued functions define some true (i.e., single-valued) functions $\tilde{v}_1, \tilde{v}_2, \tilde{w}$ on the universal cover $\widetilde{S}$ of S. Of course, $\tilde{v}_1, \tilde{v}_2, \tilde{w}$ are not arbitrary functions. In particular, since the space of the solutions of (E) is two-dimensional, two local determination of the basis (v_1, v_2) can be mapped into one another by a linear transformation. It follows that two local determinations of the multi-valued function w can be mapped into one another by a homography (i.e., a transformation of the type $w \mapsto \frac{aw+b}{cw+d}$). This means that there exists a representation ρ of the fundamental group $\pi_1(S)$ in the group $\mathrm{PSL}(2, \mathbb{C})$ of all homographies, such that the function $\widetilde{w} : \widetilde{S} \to \widehat{\mathbb{C}}$ satisfies $\widetilde{w}(\gamma \cdot \widetilde{x}) = \rho(\gamma) \cdot \widetilde{x}(\widetilde{x})$ for every $\gamma \in \pi_1(S)$. The representation ρ is called the *monodromy representation* associated with the equation (E). Up to conjugacy, ρ does not depend on the choice of v_1 and v_2.

If the range of the function $\widetilde{w} : \widetilde{S} \to \widehat{\mathbb{C}}$ happens to be contained in the unit disc $\mathbb{D}$, and if the function $\widetilde{w}$ happens to define a global biholomorphism between $\widetilde{S}$ and $\mathbb{D}$, then:

1. the inverse of $\widetilde{w}$ uniformizes the (universal cover of the) Riemann surface S,
2. the range of the monodromy representation ρ is contained in the group of homographies which preserve $\mathbb{D}$ (which can be identified with $\mathrm{PSL}(2, \mathbb{R})$), and is a Fuchsian group.

In this particular situation, Poincaré says that the normal equation (E) is *Fuchsian.*

Conversely, one can prove that, if $\widetilde{w}$ is a biholomorphism between $\widetilde{S}$ and $\mathbb{D}$, then there exists a normal equation (E) on S (which is automatically Fuchsian), such that $\widetilde{w}$ is the ratio of two solutions of (E). Hence, the problem of uniformizing Riemann surfaces can be rephrased in the following terms: *does there exists a Fuchsian equation on every Riemann surface?* Until 1883, H. Poincaré will always think about the uniformization of Riemann surfaces in those terms.

Now, we fix an integer $g \geq 2$, and we only consider closed Riemann surfaces of genus g. Following the "continuity method" strategy, Poincaré tries to prove that the set of all Riemann surfaces which carry a Fuchsian equation is open (in the moduli space of all the closed Riemann surfaces of genus g). For this purpose, Poincaré does not hesitate to consider some very sophisticated spaces, which will be rigorously studied only in the second half of the twentieth century:

1. for every Riemann surface S (or, more precisely, for every class of Riemann surfaces modulo biholomorphism), the space $\mathcal{E}(S)$ of all the normal differential equations which live on S; this is an affine space (which is nowadays identified with the space of holomorphic quadratic differentials on S);
2. the fibered bundle $\mathcal{E}_g$ whose base space is the moduli space of all closed Riemann surface of genus g, and whose fiber over a surface S is the affine space $\mathcal{E}(S)$ (this fibered space is nowadays identified with the cotangent bundle of the moduli space of closed Riemann surfaces of genus g);
3. the space $\mathcal{R}_g^{\mathbb{C}}$ of all conjugacy classes of representations of the fundamental group of a closed surface of genus g in $\mathrm{PSL}(2,\mathbb{C})$ (nowadays known as the character variety of representations of $\pi_1(S)$ in $\mathrm{PSL}(2,\mathbb{C})$).
4. the real submanifold $\mathcal{R}_g^{\mathbb{R}}$ of the complex manifold $\mathcal{R}_g^{\mathbb{C}}$, consisting of the conjugacy classes of representations whose range is contained in $\mathrm{PSL}(2,\mathbb{R})$.

Once those spaces have been (informally) defined, Poincaré considers the monodromy map

$$\mathrm{Mon} : \mathcal{E}_g \to \mathcal{R}_g^{\mathbb{C}}$$

which maps a normal equation (E) to the monodromy representation of this equation.

Let S_0 be a closed Riemann surface of genus g, which is uniformized by the unit disc. This means that there exists a biholomorphism $\widetilde{w} : \widetilde{S} \to \mathbb{D}$. As explained above, this biholomorphism is the ratio of two solutions of a Fuchsian equation (E_0) on S_0. One wishes to prove that every Riemann surface S in a neighbourhood of S_0 is also uniformized by the unit disc, or equivalently, that every Riemann surface S close enough to S_0 admits a Fuchsian equation (E_S). Actually, it is enough to prove that every Riemann surface S in a neighbourhood of S_0 admits a normal equation (E_S) with the property that the range of the monodromy representation associated to (E_S) is contained in $\mathrm{PSL}(2,\mathbb{R})$. This will be a consequence of the following lemma:

Lemma 9. *The monodromy map* $\mathrm{Mon}{:}\mathcal{E}(S_0) \to \mathcal{R}_g^{\mathbb{C}}$ *is transverse to the real submanifold* $\mathcal{R}_g^{\mathbb{R}}$ *at the point* (E_0).

This lemma means that the inverse image $\mathrm{Mon}^{-1}(\mathcal{R}_g^{\mathbb{R}})$ contains a germ of submanifold which is transverse to the fiber $\mathcal{E}(S_0)$ at (E_0). This germ of submanifold meets the fiber $\mathcal{E}(S)$ for every Riemann surface S that is close enough to S_0. In other words, every Riemann surface S in a neighbourhood S_0 admits a normal equation (E_S), such that the range of the monodromy representation associated to (E_S) is contained in $\mathrm{PSL}(2,\mathbb{R})$.

This shows that the set of uniformizable Riemann surfaces is open. For a more detailed proof, see [St-Gervais2010, Chapter VI].

It is true that Poincaré does not prove that the spaces he considers are indeed manifolds or fibered spaces. It is also true that he does not prove Lemma 9. The necessary mathematical concepts will not be developed before the middle of the 20th century. It is nonetheless true that Poincaré's statements are correct. And how could we not be totally amazed when we see Poincaré making such a masterful use of these highly sophisticated mathematical objects!

4. Uniformization of functions

In 1882, Klein and Poincaré are both convinced to have a solid proof of the uniformization theorem for algebraic Riemann surfaces. This outstanding theorem is more or less the last important result proven by Klein, whose health will suddenly collapse during the autumn 1882 ([Rowe1989]). On the contrary, Poincaré immediately starts seeking for an even more general result: he tries to uniformize *all* Riemann surfaces! And he will "almost" achieve this goal less than one year later!

Nowadays, the uniformization theorem is seen as a classification result for Riemann surfaces. As already discussed in the introduction, the viewpoint was significantly different in the XIXth century. On the one hand, the goal of uniformization was to prove that Riemann surface could be seen as parametrized curves (more precisely, the goal of uniformization was to parametrize every Riemann surface by a single-valued function). On the other hand, Riemann surfaces were considered as tools, the true issue being to understand multi-valued functions (in particular for Poincaré, whose main interest was to understand multi-valued functions which arise as solutions of linear differential equations). The paper published by Poincaré in 1883 in the *Bulletin de la Société Mathématique de France* starts with the following statement:

Theorem 10 (Fourth Uniformization Theorem: uniformization of functions, original statement. Poincaré, 1883). *Let y be an arbitrary multi-valued analytic function of a variable x. One can always find another variable z such that both x and y are uniform functions of z.*

In order to explain the precise meaning of this statement, I will translate it into more geometric terms. Let y be a germ of holomorphic function of a complex variable x. When one extends the germ y, one gets a "multi-valued function" of x. One can nevertheless construct a maximal abstract Riemann surface S such that (the extension of) y is defined as a uniform holomorphic function on S. This surface S is called *the Riemann surface associated with the germ y*. Informally, S is a branched covering over the complex plane, whose leaves are locally in one-to-one correspondence with the local determinations of the extension of the germ y (for a formal definition of this surface, see, e.g., [St-Gervais2010, Encadré II.1]). Finding a variable z such that x and y are uniform functions of z amounts to finding a uniform parametrization of the Riemann surface S by a complex variable z. In other words, it amounts to uniformizing the Riemann surface S_y. In 1883, Poincaré does not manage to find a parametrization of the surface S which is a local biholomorphism. He has to allow the derivative of the parametrization to vanish at certain points (which implies that the parametrization is not locally one-to-one). In other words, the parametrization constructed by Poincaré is not a true covering, it is only a branched covering. More precisely, Poincaré proves the following statement:

Theorem 11 (Fourth Uniformization Theorem: uniformization of functions, modern statement. Poincaré, 1883). *Let y be a germ of analytic function of a complex variable x, and S be the Riemann surface associated to this germ*[17]*. Then there exists a branched covering $\pi : U \to S$, where U is a bounded open domain of the complex plane*[18].

Poincaré's original statement (see above) follows immediately from this modern statement: in the notations of the modern statement, y and x can be seen as uniform functions defined on the surface S, and thus, as single-valued functions of a coordinate z on the complex plane containing the domain U.

To really appreciate Poincaré's boldness, and the astounding range of this "uniformization theorem for functions", one must think about the diversity of multi-valued functions and about the complexity of certain of these functions: the solutions of ordinary differential equations are good (and particularly important, in Poincaré's viewpoint) examples of such "functions". Two years before Poincaré's paper, no mathematician would have dared to dream of such a general result!

We should nevertheless mention that, if Poincaré's result constitutes a tremendous progress from the viewpoint of the theory of functions, it is a far less satisfactory result from the viewpoint of Riemann surfaces (as it would be pointed out by Hilbert, seventeen years later, in his famous address to the 1900 Congress of Mathematicians). Notice, for example, that Poincaré's result leads to parametrizing the complex plane by a bounded domain, with a parametrization which is quite wild (it would be somehow simpler to parametrize the complex plane by itself)!

I do not want to give here a detailed proved of Theorem 11 (readers interested in such a detailed proof are referred to Chapter XI of [St-Gervais2010]). I will only present two crucial ingredients used by Poincaré in this proof (the uniformization of simply connected relatively compact domains with polygonal boundary, and the construction of the universal cover of a Riemann surface), and explain briefly how Poincaré's uses these ingredients.

4.1. The invention of the universal cover

To prove his theorem, Poincaré needs to prove that the Riemann surface associated to a germ of analytic function always admits a cover which is simply connected.

[17] Poincaré's proof actually works for any abstract Riemann surface S provided that one knows that S admits a non-constant meromorphic function (at that time, it was not known that every Riemann surface admits a non-constant meromorphic function).

[18] It was only in 1900 that W. Osgood managed to prove that one could take U to be the unit disc. This is of course a consequence of Riemann's representation theorem, but there was no rigorous proof of the latter before the beginning of the twentieth century.

So, he describes a very simple and very natural construction of such a cover. This is the first apparition of the concept of *universal cover*!

Poincaré considers the Riemann surface S associated to a germ of analytic function y of a variable x. He announces that he will construct a simply connected surface $\widetilde{S}$ which covers S. He starts by saying that $\widetilde{S}$ will be a branched cover of the x-plane. He explains that the surface $\widetilde{S}$ will be completely characterized if we know when a closed loop C in the x-plane lifts as a closed loop in $\widetilde{S}$, and when it lifts as an arc with two different end points.

That being said, Poincaré distinguishes two categories of loops in the x-plane:

1. those such that the continuation of $y(x)$ does not come back to its initial value when the variable x goes around the loop C;
2. those such that the continuation of $y(x)$ comes back to its initial value when the variable x goes around the loop C.

He further subdivides the second category of loops into two types:

1. a loop is of the first type if it can be continuously deformed into a single point, in such a way that, during the deformation, the loop always remains of the second category;
2. otherwise the loop is said to be of second type.

It remains to say that a loop C in the x-plane will lift as a closed loop in $\widetilde{S}$ if and only if C is of the second category and of the first type. This completely defines the surface $\widetilde{S}$.

This construction of the universal cover of a Riemann surface, although totally elementary, can be considered as the most important point of Poincaré's paper (this is, for example, B. Osgood's opinion, see [Osgood1998]). Thanks to this construction, the mathematicians looking for the general uniformization theorem may now focus their attention on simply connected Riemann surfaces[19].

4.2. Uniformization of relatively compact domains with polygonal boundaries

Let D be an open domain in a Riemann surface S. The boundary of D is said to be *polygonal* if it is a one-dimensional submanifold of S, and if there exists a holomorphic atlas of S such that, for each chart $\phi : U \to \mathbb{C}$ of this atlas, $\phi(\partial D \cap U)$ is either a line segment, or the union of two line segments with a common end. The following result is a key ingredient of Poincaré's paper:

Theorem 12 (Schwarz, 1870). *Every simply connected relatively compact domain with polygonal boundary in a Riemann surface is biholomorphic to the unit disc.*

[19] Poincaré's construction only concerns Riemann surfaces associated to germs of analytic functions. But this construction can be generalized without changing a single word to any Riemann surface, seen as a branched cover over the Riemann sphere. Recall that the modern concept of abstract Riemann surface only dates back to Weyl's celebrated book published in 1913 ([Weyl1913]). Before that book, Riemann surfaces were *by definition* seen as branched covers over the Riemann sphere.

Poincaré attributes this result to Schwarz. Although there is no such statement in Schwarz's complete works, it is true that the *alternating method* invented by Schwarz in [Schwarz1870] (see below) allows one to prove Theorem 12.

The proof of Theorem 12 consists in constructing a *Green function.*

Definition 13. Let S be a Riemann surface and p_0 be a point of S. A function $u : S \to \mathbb{R}$ has a *simple logarithmic singularity* at p_0 if there exist a neighbourhood V of p_0 and a holomorphic chart $z : V \to \mathbb{C}$ such that the function $p \mapsto u(p) + \log|z(p) - z(p_0)|$ (which is defined on V) is bounded near p_0.

Definition 14. Let S be a Riemann surface, and Ω be an open domain in S. A *Green function* on Ω is a function $u : \Omega \to \mathbb{R}$ for which there exists a point $p_0 \in \Omega$ such that:

1. u has a simple logarithmic singularity at p_0;
2. u is harmonic on $\Omega \setminus \{p_0\}$;
3. $u(p)$ tends to 0 when p leaves every compact set of Ω.

Riemann had already understood the importance of Green's functions for the uniformization of simply connected domains:

Lemma 15 (Riemann). *Let Ω be a simply connected open domain in a Riemann surface. If Ω admits a Green function, then Ω is biholomorphic to the unit disc.*

More precisely, if g is a Green function on a simply connected open domain Ω, and g^* is the *harmonic conjugate function* of g (which means that g^* is a primitive of the 1-form $\xi \mapsto dg(i\xi)$), then the map $G := \exp(g + ig^*)$ is a biholomorphism from Ω to the unit disc.

According to Lemma 15, proving Theorem 12 amounts to proving that every simply connected relatively compact open domain with polygonal boundary admits a Green function. This can be done using the so-called *Schwarz alternating method.* I will not describe precisely this method; let me just say that it very roughly goes as follows:

1. Using the fact that Ω has a polygonal boundary, one proves that Ω can be covered by a finite number of open domains $U_1, \dots, U_q$ with the following property: for every i, there exists a biholomorphism from U_i onto the unit disc $\mathbb{D}$ which extends to a homeomorphism from $\overline{U_i}$ onto the closed unit disc $\overline{\mathbb{D}}$. The existence of this biholomorphism ϕ_i allows one to solve the Dirichlet problem on the open domain U_i: given any function $v : \partial U_i \to \mathbb{R}$, one can construct a continuous function $\overline{v} : \overline{U}_i \to \mathbb{R}$ which is harmonic on U_i, and such that $\overline{v} = v$ on ∂U_i (in order to construct such a function $\overline{v}$, one can transfer the problem to the unit disc using the biholomorphism ϕ_i, and use the so-called Poisson kernel).
2. One chooses arbitrarily a point $p_0 \in \Omega$ and a function $u_0 : \Omega \to \mathbb{R}$ with a simple logarithmic singularity at p_0. Then, one constructs a sequence of functions $(u_n)_{n \geq 0}$ as follows: if n is equal to r modulo q, then u_n is a continuous

function which coincides with u_{n-1} on $\Omega \setminus U_r$ and is harmonic on U_r (except when $p_0 \in U_r$, in which case u_n is a continuous function which coincides with u_{n-1} on $\Omega \setminus U_r$, is harmonic on $U_r - \{p_0\}$, and has a simple logarithmic singularity at p_0). Notice that the u_n's are harmonic alternatively on the domains $U_1, \ldots, U_q$.

3. One proves that the sequence $(u_n)_{n\geq 0}$ converges (this is of course the delicate point; it relies on a tricky use of the maximum principle). Once the convergence is proven, it is easy to check that the limit u of the sequence $(u_n)_{n\geq 0}$ is a Green function for Ω.

For more details on this construction, see for example [St-Gervais2010, Part XI.1].

4.3. Poincaré's strategy to prove Theorem 11

Let y be a germ of analytic function of a variable x, and S be the Riemann surface associated with this germ. Recall that S is naturally a branched cover over the x-plane. Therefore, the variable x can be seen as a non-constant meromorphic function on the surface S.

The first step of Poincaré's proof is very tricky. He constructs a Riemann surface Σ which is a branched cover of S, and such that there exists a non-constant holomorphic function h from Σ into the unit disc. Actually, he constructs a Riemann surface Σ which is both a branched cover of S, and a branched cover of the unit disc (the function h is the projection of Σ onto the unit disc). This construction uses the meromorphic function $x : S \to \widehat{\mathbb{C}}$ and a Fuchsian function[20] $F : \mathbb{D} \to \widehat{\mathbb{C}}$.

Then one constructs the universal cover $\widetilde{\Sigma}$ of the Riemann surface Σ, as explained in Subsection 4.1. Since Σ is a branched cover of S, proving Theorem 11 reduces to proving that $\widetilde{\Sigma}$ is biholomorphic to a bounded open domain of the complex plane. This is precisely what Poincaré will do.

Poincaré considers an exhaustion of $\widetilde{\Sigma}$ by an increasing sequence of simply connected relatively compact open domains $D_0 \subset D_1 \subset D_2 \subset \ldots$ with polygonal boundaries, and chooses a point $p_0 \in D_0$. We have seen that, for every k, there exists a Green function $g_k : D_k \to \mathbb{R}$ which vanishes on the boundary of D_k, has a simple logarithmic singularity at p_0, and is harmonic on $D_k - \{p_0\}$.

In order to prove the convergence of the sequence (g_k), the key point is the following: if we consider a lift $\widetilde{h} : \widetilde{\Sigma} \to \mathbb{D}$ of the map $h : \Sigma \to \mathbb{D}$, and the function $t := -\log|\widetilde{h}|$, then the functions g_k are bounded from above by t. Indeed, $t - g_k$ is positive on ∂D_k (since g_k vanishes on ∂D_k, and since $t := -\log|\widetilde{h}|$ is positive on ∂D_k), and therefore it is positive on D_k by the maximum principle.

Hence, the sequence $(g_k(p))$ is bounded for every point p which is not a singularity of $\widetilde{h}$. Since (g_k) is an increasing sequence, this implies that (g_k) converges

[20] Recall that a Fuchsian function is a meromorphic function $F : \mathbb{D} \to \widehat{\mathbb{C}}$ which is invariant under a lattice Γ of $\mathrm{Auto}(\mathbb{D}) \simeq \mathrm{PSL}(2, \mathbb{R})$. Also recall that the existence of such functions had been discovered by Poincaré two years before.

pointwise to a function $g : \widetilde{\Sigma} \to [0, +\infty]$ which is finite at every point p which is not a singularity of $\widetilde{h}$. Now, using the Dominated Convergence Theorem, it is easy to prove that g has a simple logarithmic singularity at p_0, and is harmonic on $\widetilde{\Sigma} - \{p_0\}$.

For every k, the map $G_k := \exp(g_k + ig_k^*)$ is a biholomorphism from D_k onto the unit disc $\mathbb{D}$ (see Lemma 15). Since the sequence of functions (g_k) converges to the function g, the sequence of biholomorphisms (G_k) converges (uniformly on compact sets) to the holomorphic map $G := \exp(g + ig^*)$. Since g has a simple logarithmic singularity, G is not constant. Moreover, g is one-to-one, since this is the case for the g_k's. Therefore, G is a biholomorphism from $\widetilde{\Sigma}$ onto an open domain of the unit disc, and the proof of Theorem 11 is complete. For more details, see [St-Gervais2010, Part XI.3].

Remark 16. A more naive strategy would be to consider the universal cover $\widetilde{S}$ of the Riemann surface S, and an exhaustion[21] of $\widetilde{S}$ by simply connected relatively compact open domains with polygonal boundaries $D_0 \subset D_1 \subset D_2 \subset \cdots$. Each domain D_k would admit a Green function g_k, and the function $G_k := \exp(g_k + ig_k^*)$ would define a biholomorphism from D_k onto the unit disc. The sequence of Green's function $(g_k)_{k\geq 0}$ would of course be increasing. But the problem is that there is no guarantee that this sequence will be bounded (actually, there are some situations where this sequence is *not* bounded!). So this naive strategy fails.

5. Solving the Liouville equation: an alternative method for uniformizing algebraic Riemann surfaces

I do not know who (was it H. Schwarz?) has first noticed that the uniformization of Riemann surfaces was intimately linked with solving of the Liouville equation

$$\Delta u = 2e^u.$$

What is for sure is that this link is explicitly mentioned in the statement of problem submitted to the mathematical community in 1890 by the *Royal Sciences Society of Göttingen.*

> *The problem of the conformal representation of a planar domain* [i.e., *a domain in the complex plane, or a domain in a Riemann surfaces which is seen as a branched cover of the complex plane*] *or a portion of a* [*Riemannian*] *surface of constant curvature k is linked to the problem of the solving of the partial differential equation*
>
> $$\Delta u = \frac{\partial^2 u}{\partial x^2} + \frac{\partial^2 u}{\partial y^2} = -2ke^u$$
>
> *with prescribed boundary values and prescribed singularities.*

[21] Let us leave apart the case where $\widetilde{S}$ is homeomorphic to the sphere, which had been treated by Schwarz in 1870; see of S by [Schwarz1870] or [St-Gervais2010, Part IV.1].

For this last problem, one should first consider the boundary values and the singularity types that were specified by Riemann in his theory of Abelian functions.

The Royal Society wishes to obtain a complete answer to the following question: is it possible to solve the above differential equation on a given domain, with some prescribed boundary values and some prescribed types of singularities, under the hypothesis that the constant k is negative?

In particular, the Royal Society wishes the question to be treated in the case where the planar domain under consideration is a closed Riemann surface with several leaves, and where the function u is required to have only logarithmic singularities.

Émile Picard published a solution to this problem before the end of the year 1890 ([Picard1890]). Yet, in this 1890's paper, Picard only considered the question of the existence of solutions to the Liouville equation $\Delta u = -2e^u$ (with prescribed singularities) on a bounded domain of $\mathbb{C}$. He claimed that the case of a general Riemann surface did not lead any additional difficulty... but he soon realized that this was an excessive assertion. During the following 15 years, he felt the need to come back to the problem in several papers (in 1893 [Picard1893c, Picard1893b, Picard1893a], in 1898 [Picard1898], and in 1905 [Picard1900]), in order to repair his proof, and to explain how to generalize it to the case of a closed Riemann surface. For a detailed exposition of Picard's method for solving the Liouville equation, see [Picard1931, Chapitre 4].

In 1898, Poincaré published a memoir containing his own solution to the problem raised by the Royal Society of Sciences of Göttingen.

One of the particularly interesting features of this memoir is that Poincaré adopts an "intrinsic" viewpoint. The Liouville equation $\Delta u = ke^u$ is only meaningful on a domain of $\mathbb{C}$, or on a Riemann surface which is explicitly seen as a branched cover of $\mathbb{C}$ (indeed, one needs some coordinates to define the Laplace operator Δ). Therefore, in order to relate the Liouville equation on a Riemann surface S to the existence of a uniformization of S, one needs to choose a meromorphic function f on S which will allow S to be seen as a branched cover of $\widehat{\mathbb{C}}$, and to seek solutions of the Liouville equation which have singularities of a given type at the zeroes and the poles of f. Notice nevertheless that these singularities are "artificial", in the sense that the surface S itself is not singular. In his memoir, Poincaré constructs a "Riemannian Liouville equation" directly on a Riemann surface S, without needing a meromorphic coordinate (see below).

The results contained in the memoir of Poincaré provide a proof of the Uniformization Theorem for algebraic Riemann surfaces. In our opinion, this is the first rigorous proof of this result: indeed, the "proofs" of Klein and Poincaré based on the continuity method (see Section 3) are far from being totally convincing, and we consider that the resolution of the Liouville equation by Picard was completed (for closed Riemann surfaces) only in 1905.

Before presenting Poincaré's strategy to solve the Liouville equation, let us explain why solving the Liouville equation leads to the uniformization of Riemann surfaces.

5.1. Why solving the Liouville equation leads to the Uniformization Theorem

5.1.1. From a uniformizing function to a solution of the Liouville equation. Let us first consider a simply connected domain U of the complex plane. Suppose that U is uniformized by the unit disc $\mathbb{D}$. Since $\mathbb{D}$ is simply connected, this means that there exists a global biholomorphism $f : \mathbb{D} \to U$. Using f, we can transfer the hyperbolic metric of $\mathbb{D}$ to U: we obtain a Riemannian metric g_{hyp} on U given by the formula

$$g_{\mathrm{hyp}} = 4\frac{\left|\frac{df^{-1}}{dz}\right|^2}{(1-|f^{-1}|^2)^2}dzd\overline{z}.$$

Now let us consider the conformal factor

$$e^{u_z} = 4\frac{\left|\frac{df^{-1}}{dz}\right|^2}{(1-|f^{-1}|^2)^2}$$

which relates the metric g_{hyp} to the Euclidean metric $dzd\overline{z}$. Using the fact that f^{-1} is a holomorphic function, one gets

$$\frac{\partial^2 u_z}{\partial z \partial \overline{z}} = \frac{2\frac{df^{-1}}{dz}\frac{d\overline{f^{-1}}}{d\overline{z}}}{\left(1 - f^{-1}\overline{f^{-1}}\right)^2} = \frac{1}{2}e^{u_z}.$$

Hence, the function $u_z : U \to \mathbb{R}$ is a solution of the Liouville equation

$$\Delta_z u = 2e^u, \tag{3}$$

where $\Delta_z = 4\frac{\partial^2}{\partial z \partial \overline{z}}$ is the usual Laplace operator on the complex plane. Therefore, we have proved the following fact: *if U is a simply connected domain of the complex plane, and U is uniformized by the unit disc, then the Liouville equation* (3) *admits a solution defined on U.*

Now, we consider a Riemann surface S. We assume that S is uniformized by the unit disc $\mathbb{D}$, i.e., we assume that there exists a covering map $f : \mathbb{D} \to S$ which is a local biholomorphism. As in the preceding paragraph, we may use f to transfer the hyperbolic metric of $\mathbb{D}$ to S: indeed, f is locally invertible, and therefore admits a multi-valued inverse f^{-1}; two local determinations of f^{-1} differ by a biholomorphism of $\mathbb{D}$, and such an biholomorphism preserves the hyperbolic metric of $\mathbb{D}$. We obtain a Riemannian metric g_{hyp} on the surface S. Now, let us choose a meromorphic function z on S, which allows us to see S as a branched cover over (an open set of) the Riemann sphere $\widehat{\mathbb{C}}$. We can regard z as a coordinate on $\mathbb{C} = \widehat{\mathbb{C}} - \{\infty\}$. As in the previous paragraph, we have

$$g_{\mathrm{hyp}} = 4\frac{\left|\frac{df^{-1}}{dz}\right|^2}{(1-|f^{-1}|^2)^2}dzd\overline{z}.$$

Observe nevertheless that the derivative $\frac{df^{-1}}{dz}$ is infinite at branching points of z and vanishes at the poles of z. We can define a function $u_z : S \to \mathbb{R}$ just as in the previous paragraph; this will again be a solution of the Liouville equation (3). But this function u_z will exhibit singularities at the branching points and poles of z. The types of singularities that occur can easily be described (for example, at branching points, the singularities are logarithmic, with a coefficient that depends only on the branching order). So we have the following result: *if a Riemann surface S is uniformized by the unit disc, then the Liouville equation* (3) *admits a solution with prescribed singularities.*

As already noticed, the link described above between the uniformization of a Riemann surface S and the Liouville equation is not very satisfactory. Indeed, it introduces some artificial singularities (notice that, in the previous paragraph, the metric g_{hyp} is not singular; the singularities come the "meromorphic chart z" which maps the Riemann surface S on the complex plane). Actually, the problem comes from the fact that one tries to use a differential operator (the Laplace operator Δ_z) which is not well defined on S.

In his memoir, Poincaré adapts an innovative point of view, more intrinsic and more natural. He shows that one can construct a "Laplace operator" on the surface S itself, and therefore replace equation (3) by an "Riemannian Liouville equation" which does not depend of a particular coordinate system on S. The uniformizing functions for S now correspond to non-singular solutions of this "Riemannian Liouville equation".

Let us explain this. Let S be a Riemann surface. To every local holomorphic chart $z : U_z \subset S \to \mathbb{C}$ is associated a Laplace operator $\Delta_z = 4\frac{\partial}{\partial z \partial \bar{z}}$, and a Liouville equation $\Delta_z u = -2e^u$ (which of course is defined only on the domain U_z of the chart z). Poincaré shows that the Liouville equations associated to an atlas of holomorphic charts can be "glued together" to produce a "*global* Liouville equation" on the surface S. For this purpose, he considers a Riemannian metric g on S, which he assumes to be compatible with the complex structure (i.e., the conformal structure defined by g coincides with that defined by the complex structure). For every local holomorphic chart $z : U_z \subset S \to \mathbb{C}$, there exists a function $\sigma_z : U_z \to \mathbb{R}$ such that $g = e^{-\sigma_z} dz d\bar{z}$. Poincaré first shows that the metric g allows one to "glue together" the Laplace operators defined by the various charts. In other words, he shows that there exists a differential operator $\Delta_g : C^2(S, \mathbb{R}) \to C^0(S, \mathbb{R})$ (nowadays called the *Laplace–Beltrami operator*) such that, for every local holomorphic chart $z : U_z \subset S \to \mathbb{C}$,

$$\Delta_g = 4e^{\sigma_z} \Delta_z$$

on the domain U_z. Then, Poincaré shows that one can construct a "Riemannian Liouville equation". More precisely, he proves that a function $u : S \to \mathbb{R}$ is a solution of the Liouville equation $\Delta_z u = 2e^u$ on the domain U_z for every local holomorphic chart $z : U \subset S \to \mathbb{C}$, if and only if u satisfies the equation

$$\Delta_g = 2e^u - \phi_g, \tag{4}$$

where $\phi_g : S \to \mathbb{R}$ is the function defined by $\phi_g = -\frac{1}{2}\Delta_g\sigma_z$ in each local chart z. The same arguments as in the previous paragraph show that:

Proposition 17. *If the surface S is uniformized by the unit disc, then the "Riemannian Liouville equation" equation* (4) *admits a (non-singular) solution.*

5.1.2. From a solution of the Liouville equation to a uniformizing function. In his memoir, Poincaré explains – just as we did above – why the existence of a uniformizing function implies the existence of a solution to the Liouville equation (4). Quite surprisingly, he does not state the converse (although he seems to be perfectly aware that this converse does hold[22],[23]). The proof of this converse goes through the interpretation of the Riemannian Liouville equation (4) in terms of curvature.

Let U be domain of the complex plane, and $u : U \to \mathbb{R}$ be a C^2 function. A (painful, but) elementary calculation shows that the Gauss curvature of the Riemannian metric $e^u dzd\overline{z}$ is equal to

$$-\frac{1}{2}e^{-u}\Delta_z u \tag{5}$$

(see for example [Jost2002]). One easily deduces that a function u is a solution of the Liouville equation (3) if and only if the Riemannian metric $e^u dzd\overline{z}$ has constant curvature -1.

Now let S be a closed Riemann surface with negative Euler characteristic, and g be a Riemannian metric on S compatible with the complex structure of S. We consider the function $\phi_g : S \to \mathbb{R}$ defined as in Subsection 5.1.1: for every holomorphic chart z, one has $\phi_g = -\frac{1}{2}\Delta_g\sigma_z = -2e^{\sigma_z}\Delta_z\sigma_z$, where σ_z is the function such that $g = e^{-\sigma_z}dzd\overline{z}$. Formula (5) shows that ϕ_g is nothing but the Gauss curvature of the metric g with the minus sign. Using once again formula (5), we deduce that the function u is a solution of the Riemannian Liouville equation (4) if and only if the metric $e^u g$ has constant curvature -1.

So, if the Riemannian Liouville equation (4) admits a solution u, then the universal covering space of S is a simply connected complete Riemannian surface with constant Gauss curvature -1. It is well known that such a Riemannian surface is isometric to the hyperbolic disc. This proves the following:

---

[22]He indeed writes

> *The integration of* [the Liouville equation (4)] *would lead directly to the solution of the problem which preoccupies us* [the existence of uniformizing functions for algebraic Riemann surfaces].

[23]Poincaré considered that he had proved the Uniformization Theorem for algebraic Riemann surfaces fifteen years before, thanks to the continuity method. Therefore the aim of Poincaré was not really to provide another proof of the Uniformization Theorem, but rather to develop some tools to *calculate* the uniformizing function of a given complex algebraic curve. For example, although he does not prove (nor even state) that solving the Liouville equation yields a new proof of the Uniformization Theorem, he carefully explains how to recover a uniformizing function (assumed to exist) from a solution of the Riemannian Liouville equation (4).

Proposition 18. *If the Riemannian Liouville equation* (4) *admits a solution, then the surface S can be uniformized by the unit disc.*

5.2. How Poincaré solved the Liouville equation

Let S be a closed (hence algebraic) Riemann surface[24],[25] with negative Euler characteristic. We endow S with a Riemannian metric g compatible with the complex structure of S. As explained before, proving that S is uniformized by the disc amounts to proving that the Riemannian Liouville equation (4) admits a solution u. In his memoir, Poincaré actually proves the existence and uniqueness of a solution for the more general equation

$$\Delta_g u = \theta e^u - \phi, \tag{6}$$

where $\theta : S \to \mathbb{R}$ and $\phi : S \to \mathbb{R}$ are arbitrary functions with positive mean values. We shall now explain how Poincaré manages to solve this equation (which we shall call the *generalized Riemannian Liouville equation.*

Poincaré's strategy can be summarized as follows. He starts with a partial differential equation on S that can be solved explicitly: the Poisson equation $\Delta_g u = -\psi$ (for an arbitrary function ψ with zero mean value). Then he tries to solve partial differential equations "in a neighbourhood" of a Poisson equation, by showing that the solutions of these equations can be expressed as series of functions, each of which satisfies a Poisson equation. Then, he repeats the same procedure. At each step, the set of PDEs that can be solved becomes larger and larger. Of course, the goal is to reach the generalized Riemannian Liouville equation (6) after a finite number of steps.

It might be interesting to notice the analogy between this strategy and the *continuity method* (see Section 3). Here, Poincaré starts with a PDE which he can solve explicitly, and makes his way in the realm of all PDEs on S, solving more and more equations, hoping to reach the generalized Riemannian Liouville equation (6) after a finite number of steps. The continuity method consisted in starting with a given Riemann surface which could be uniformized explicitly, and making one's way in the realm of all Riemann surfaces, hoping to reach the Riemann surface S.

Let us give more details on Poincaré's proof.

1. He first considers the Poisson equation $\Delta_g u = -\psi$, where ψ is a function with zero mean value. An explicit solution of this equation had been known for almost a century in the particular case where S is the complex plane:

[24]Poincaré uses the compactness of S to ensure that certain functions defined on S are bounded. He also uses the fact that S is algebraic (it has be known since the seminal work of Riemann that a closed Riemann surface "is" algebraic) to construct meromorphic function with prescribed poles on S: at that time, the Dirichlet problem for the Laplacian was not solved; therefore the only way to construct meromorphic functions on a given Riemann surface S was to define "explicitly" these functions using an algebraic equation of S and some *Abelian integrals*; see, e.g., [St-Gervais2010, II.2.3]).

[25]Actually, Poincaré also considers Riemann surfaces with punctures, but we were not able to understand what precise statement he actually proves in this context.

indeed, the function $p \mapsto \int \log |p-q|\psi(q)dv_g(q)$ satisfies the equation (where dv_g is the area form associated to g). In other words, on the complex plane, the solution of the Poisson equation is obtained as the convolution of the function ψ with the *Green function* $z \mapsto \log |z|$. Poincaré generalizes this classical result. Since a compact surface cannot admit a "true Green function" (i.e., a harmonic function with a single simple logarithmic pole), he has to use a function with two poles, which leads some technical difficulties. More precisely:

(a) Since the Riemann surface S is algebraic, the theory of Abelian integrals allows to construct meromorphic functions with prescribed poles on S. For example, given an algebraic equation $F(x,y) = 0$ of S, and three pairwise distinct points p_0, q_0, q on S, elementary arguments of linear algebra show that there exist a polynomial P and three complex numbers a, b, c such that the formula

$$(x,y) \mapsto \int_{p_0}^{p} \frac{P(x,y)}{(ax+by+c)\frac{\partial F}{\partial y}(x,y)} dx$$

defines a meromorphic function on S that vanishes at p_0, has simple poles at q and q_0, and has no other poles.

(b) The real part of the primitive of this function provides a function $p \mapsto G_{p_0,q_0,q}(p)$ that vanishes at p_0, has simple logarithmic singularities at q_0 and q, and is harmonic on $S - \{q_0, q\}$.

(c) Then one can check that the function $\int_S G_{p_0,q_0,q}(p)\psi(q)\ d\mathrm{vol}_g(q)$ is a solution of the Poisson equation $\Delta_g u = -\psi$.

2. Having solved the Poisson equation, Poincaré considers the equation $\Delta_g u = \eta u - \phi$.

(a) He first proves that the equation $\Delta_g u = \lambda \eta u - \phi$ can be solved for any given functions η and ϕ, provided that the real number λ is close enough to zero. For this purpose, he seeks for a solution under the form $u := u_0 + \lambda u_1 + \lambda^2 u_2 + \cdots$. The crucial observation is the following: if u satisfies the equation $\Delta_g u = \lambda \eta u - \phi$, then the functions $u_0, u_1, u_2, \dots$ must satisfy some Poisson equations. This is a very interesting information, since Poincaré has previously proved that every Poisson equation admits a (unique) solution. Now Poincaré shows that, conversely, if $u_0, u_1, u_2, \dots$ satisfy some Poisson equations, and λ is small enough, then the series $u_0 + \lambda u_1 + \lambda^2 u_2 + \cdots$ converges, and the sum of the series is a solution of $\Delta_g u = \lambda \eta u - \phi$.

(b) Using another series expansion, Poincaré proves that, if one can solve the equation $\Delta_g u = \lambda_0 \eta u - \phi$ for a given λ_0, then one can also solve the equation $\Delta_g u = (\lambda_0 + \lambda)\eta u - \phi$ provided that $\lambda < \lambda_0$.

(c) From the two preceding items, it immediately follows that the equation $\Delta_g u = \lambda \eta u - \phi$ admits a solution for every $\lambda > 0$. In particular, the equation $\Delta_g u = \eta u - \phi$ admits a (unique) solution.

3. Poincaré finally settles the equation $\Delta_g u = \theta e^u - \phi$.

(a) He first observes that $\Delta_g u = \theta e^u - \phi$ admits a trivial (constant) solution in the particular case where ϕ is proportional to θ.

(b) Them he shows (using once again a series expansion) that, if one can solve $\Delta_g u = \theta e^u - \phi_0$ for a given function ϕ_0, then one can also solve $\Delta_g u = \theta e^u - (\phi_0 + \lambda\psi)$ for every function ψ, provided that λ is small enough. Here, the arguments that are necessary to prove the convergence of the series are very sophisticated (this is not surprising, since the equations under consideration are highly non-linear).

(c) From the two preceding items, one easily deduces that the equation $\Delta_g u = \theta e^u - \phi$ can be solved provided that the function ϕ is positive.

(d) Finally, an elementary trick allows to reduce the general case (where ϕ is any function with positive mean value) to the particular case where ϕ is a positive function.

At the end of day, Poincaré has proved the following result:

Theorem 19 (Fifth Uniformization Theorem: uniformization of closed Riemann surfaces, via the Liouville equation. Poincaré, 1898). *Let S be a closed Riemann surface with negative Euler characteristic, and g be a Riemannian metric on S which is compatible with the complex structure. For every positive C^1 function $\theta : S \to \mathbb{R}$ and every C^1 function $\phi : S \to \mathbb{R}$ with positive mean value, the generalized Riemannian Liouville equation* (6) *admits a (unique) C^2 solution $u : S \to \mathbb{R}$. As a further consequence, the Riemann surface S is uniformized by the unit disc.*

The method invented by Poincaré to solve the Liouville equation might seem a bit laborious. This method has nevertheless the advantage of using only very elementary tools. Observe that, even with all the sophisticated tools developed along the first half of the twentieth century (distributions, weak-$\star$ topology, compact embeddings of Sobolev spaces, elliptic regularity,...), solving equation (6) is still not a easy task (once again, one should not forget that this equation is nonlinear). Indeed, it was only in year 1971 that M. Berger published a modern proof of the existence and uniqueness of a solution for this equation! (see [Berger1971]).

It is interesting to notice that the existence of a solution for the equation $\Delta_g u = \theta e^u - \phi$ amounts to the existence of a Riemannian metric on S (in the conformal class of g) with Gauss curvature $-\frac{1}{2}\theta$. Therefore, Poincaré's memoir provides a complete solution to the problem of the existence of Riemannian metrics with prescribed variable Gauss curvature on closed surfaces. This problem was considered to be opened in Berger's paper [Berger1971] in 1971!

A much more detailed account of Poincaré's work on the Liouville equation can be found in Chapter X of [St-Gervais2010].

6. The "sweeping method": a physical proof of the uniformization theorem

In his famous address at the International Congress of Mathematicians in 1900, Hilbert recalls that the solution proposed by Poincaré in 1883 for the uniformization of non-algebraic Riemann surfaces (see Section 4) is not fully satisfactory. He invites geometers to look again at this problem. The uniformization of Riemann surfaces is the 22nd of the famous 23 Hilbert's problems.

> *As Poincaré was the first to prove, it is always possible to reduce any algebraic relation between two variables to uniformity by the use of automorphic functions of one variable. That is, if any algebraic equation in two variables be given, there can always be found for these variables two such single-valued automorphic functions of a single variable that their substitution renders the given algebraic equation an identity. The generalization of this fundamental theorem to any analytic non-algebraic relations whatever between two variables has likewise been attempted with success by Poincaré, though by a way entirely different from that which served him in the special problem first mentioned. From Poincaré's proof of the possibility of reducing to uniformity an arbitrary analytic relation between two variables, however, it does not become apparent whether the resolving functions can be determined to meet certain additional conditions. Namely, it is not shown whether the two single-valued functions of the one new variable can be so chosen that, while this variable traverses the regular domain of those functions, the totality of all regular points of the given analytic field are actually reached and represented. On the contrary it seems to be the case, from Poincaré's investigations, that there are beside the branch points certain others, in general infinitely many other discrete exceptional points of the analytic field, that can be reached only by making the new variable approach certain limiting points of the functions. In view of the fundamental importance of Poincaré's formulation of the question it seems to me that an elucidation and resolution of this difficulty is extremely desirable.*
>
> *In conjunction with this problem comes up the problem of reducing to uniformity an algebraic or any other analytic relation among three or more complex variables, a problem which is known to be solvable in many particular cases. Toward the solution of this the recent investigations of Picard on algebraic functions of two variables are to be regarded as welcome and important preliminary studies.*

The uniformization of Riemann surfaces will be one of the first Hilbert's problem to be solved. In 1907, Koebe and Poincaré published simultaneously independent (and quite different from each other) proofs of the statement that is known today as *the Uniformization Theorem* ([Koebe1907a, Poincaré1907]):

Theorem 20 (Sixth Uniformization Theorem: Poincaré–Koebe Uniformization Theorem. 1907). *Every simply connected Riemann surface*[26] *is biholomorphic to the Riemann sphere, the complex plane, or the unit disc.*

Koebe's proof is nice and totally correct, but we will not even sketch it here since the present text focuses on Poincaré's contributions (for an account of Koebe's proof, see [St-Gervais2010, Chapter XII]). On the contrary, I will describe Poincaré's proof quite in details. This proof is rather original, since it consists in a mathematical translation of an electrostatic experiment (or rather a electrostatic gedankenexperiment)[27].

6.1. Poincaré's strategy

Let S be a simply connected Riemann surface. Schwarz had already proved in 1870 that the Riemann sphere is the only compact simply connected Riemann surface (up to biholomorphism). So we may, and we will, assume that S is non-compact. Poincaré's proof uses a result that was proved by Osgood a few years before ([Osgood1900]). In order to state Osgood's result, I first need to introduce a generalization of the concept of Green's function:

Definition 21. Let S be a Riemann surface. A *generalized Green's function* is a positive function $f : S \to (0, +\infty)$ with a discrete set of singularities, such that:

- f is which is harmonic outside its set of singularities;
- $f(x)$ tends to $+\infty$ when x approaches a singularity;
- there is at least one singularity p_0 which is a simple logarithmic singularity (i.e., given a local holomorphic coordinate z on a neighbourhood of p_0, the function $f(p) + \log|z(p) - z(p_0)|$ is bounded near p_0).

Theorem 22 (Osgood, 1900). *If a Riemann surface S carries a generalized Green function, then the universal cover of S is biholomorphic to the unit disc.*

The proof of Osgood's theorem is quite tricky[28], but essentially relies on the same strategy as the proof of Poincaré's fourth uniformization theorem presented in Section 4 (recall that this strategy roughly consists in ensuring the convergence of the sequence of Green's functions associated with an exhaustion of the universal cover of S by relatively compact discs with polygonal boundaries).

Poincaré does not try to construct directly a generalized Green function on the surface S: he "makes a small hole" in S to get a surface with boundary. This boundary will play a crucial role in the proof: it will allow to control the growth of a sequence of functions that will converge to a generalized Green function.

---

[26]Both Koebe and Poincaré's proofs require the existence of a meromorphic function on the Riemann surface under consideration. The existence of a meromorphic function is actually automatic, but this is not a trivial fact. At that time, all the Riemann surfaces were *by definition* assumed to be branched covers over the Riemann sphere; such a Riemann surface of course admits a meromorphic function.

[27]Actually, Poincaré does not even make a complete "mathematical translation": some statements are considered by him as "physical evidence".

[28]In fact, Theorem 22 is even named "Osgood's trick" by some authors.

More precisely, we pick a local holomorphic chart z which maps an open set of S onto a neighbourhood of 0 in $\mathbb{C}$. We choose a real number $r > 0$ such that the closed disc $\overline{\mathbb{D}}(0,r)$ is contained in the range of z, and we set

$$A := S \setminus z^{-1}(\overline{\mathbb{D}}(0,r)).$$

Notice that A is an open domain in the surface S which is homeomorphic to an annulus, and that the boundary of A in S is the curve $\partial A = \{p \in S\ ;\ |z(p)| = r\}$. The following proposition follows directly from Osgood's Theorem 22 (see [St-Gervais2010, Proposition XII.1.1].

Proposition 23. *Suppose that the Riemann surface A admits a generalized Green function. Then S is biholomorphic to the complex plane or to the unit disc.*

6.2. Existence of a generalized Green function on A

According to Proposition 23, it remains to prove the existence of a generalized Green function on the Riemann surface A. For this purpose, Poincaré uses a method based on an electrostatic analogy, which he calls the *sweeping* (balayage) *method.*

We seek for a generalized Green function with a single singularity. Let us recall that this is a positive function $u : A \to \mathbb{R}$, which has a simple logarithmic singularity at some point $p_0 \in A$ and is harmonic on $A \setminus \{p_0\}$. Poincaré regards functions on A as *electrostatic potentials.* According to the laws of electrostatics[29], a potential $v : A \to \mathbb{R}$ is created by an electric charge distribution of density Δv (where Δ is the Laplace operator). Therefore, a generalized Green function $u : A \to \mathbb{R}$ with a simple logarithmic singularity at p_0 would correspond to a electric charge distribution consisting of a single positive point charge located at p_0.

Poincaré starts with an arbitrary positive function $u_0 : A \to \mathbb{R}$ with a simple logarithmic singularity at p_0. The electric charge distribution corresponding to u_0 consists of

- a positive point charge at p_0, and
- a distribution of negative charges on $A \setminus \{p_0\}$.

In order to "transform u_0 into a generalized Green function", one needs to "sweep the negative charges towards infinity". For this purpose, Poincaré imagines that "some parts of A suddenly become electrical conductors". From a mathematical viewpoint, "making a domain D of A suddenly become an electrical conductor" amounts to replacing the electrostatic potential v by an electrostatic potential v' that coincides with v outside D, and is harmonic inside D. Poincaré will cover A by a countable collection of holomorphic discs $D_1, D_2, \ldots,$ and "make these discs electrical conductors one after the other". This will produce a sequence of electrostatic potentials $(u_n)_{n\geq 0}$, which is hoped to converge towards a generalized Green function.

[29] These laws have no true physical meaning on A, since A is an abstract Riemann surface... but we will only use these laws as kind of guides for our intuition.

Let us now describe the *sweeping method* from a more mathematical viewpoint. We fix a point p_0 in A. Recall that a function v has a *simple logarithmic singularity at* p_0 if, given a local holomorphic coordinate z defined on a neighbourhood of p_0, the function $v(p) + \log|z(p) - z(p_0)|$ is bounded near p_0.

Let D be an open domain of A, such that there exists a biholomorphism from D to the unit disc which extends as a homeomorphism from $\overline{D}$ to the closed unit disc (in particular, D is relatively compact). Given a continuous function $w : \partial D \to \mathbb{R}$, there exists a unique function $\overline{w} : \overline{D} \to \mathbb{R}$ which is continuous and harmonic on D. This function $\overline{w}$ is called the *harmonic extension* on w on D. It can be obtained as the convolution of w with the Poisson kernel (after having mapped $\overline{D}$ to the closed unit disc in the complex plane). Moreover, if p_0 is in D, there exists a unique Green function on D with singularity at p_0, i.e., there exists a unique function $\overline{w} : \overline{D} \to \mathbb{R}$, which is continuous on $\overline{D} - \{p_0\}$, is harmonic on $D - \{p_0\}$, vanishes on ∂D, and has a simple logarithmic singularity at p_0.

Now, let v be a continuous real-valued function defined on $A \setminus \{p_0\}$, with a simple logarithmic singularity at p_0. We denote by v' the continuous function which coincides with v outside D, and is defined as follows on D:

- if p_0 is not in D, then $v'_{|D}$ is the harmonic extension of $v_{|\partial D}$;
- if p_0 is in D, then $v'_{|D}$ is the sum of the harmonic extension of $v_{|\partial D}$ and of the Green function of D with singularity at p_0.

Observe that v' has a simple logarithmic singularity at p_0 in both cases. We say that v' is obtained from v by *sweeping electric charges outside* D.

The sweeping method consists in repeating this operation infinitely many times on discs that cover the surface A. More precisely, we consider a countable family of open domains $\mathcal{D} = \{D_1, D_2, D_3, \dots\}$, such that:

- $D_1, D_2, D_3, \dots$ cover A;
- p_0 is not on the boundary of any of the D_i's,
- for every i, there exists a biholomorphism from D_i to the unit disc $\mathbb{D}$ which extends to a homeomorphism from $\overline{D_i}$ to the closed unit disc $\overline{\mathbb{D}}$.

Now we consider a sequence $(D_{i_k})_{k>0}$ of elements of the family $\mathcal{D}$ with the following property: each element of $\mathcal{D}$ appears infinitely many times in the sequence $(D_{i_k})_{k\geq 0}$. For example, we may take $D_{i_1}, D_{i_2}, D_{i_3}, \dots = D_1, D_2, D_1, D_2, D_3, D_1, D_2, D_3, D_4, \dots$. Starting from a positive continuous function $u_0 : A \setminus \{p_0\} \to \mathbb{R}$ with a simple logarithmic singularity at p_0, we define a sequence of functions $(u_n)_{n\geq 0}$ as follows: for every n, the function u_{n+1} is obtained from u_n by sweeping the electric charges outside D_{i_n}.

It remains to prove that this sequence (u_n) converges to a generalized Green function.

6.3. Convergence of the sweeping process

In order to ensure the convergence of the sequence $(u_n)_{n\geq 0}$ towards a generalized Green function, we still have the freedom of choosing the initial "electrostatic potential" u_0 (the only constraint is that u_0 should have a simple logarithmic singularity at p_0, and no other singularity). Poincaré considers a potential u_0 which is *sub-harmonic* on $A \setminus \{p_0\}$ (see the definition below); this will allow him to use the so-called *Harnack principle* (Proposition 27), which will guarantee the convergence of the sequence (u_n) provided that he can prove that this sequence does not converge to infinity at every point of A.

Definition 24. Let Σ be a Riemann surface. A continuous function $v : \Sigma \to \mathbb{R}$ is said to be *sub-harmonic* if its value at every point $p \in \Sigma$ is less than the mean of its values on a circle centered at p. More formally, $v : \Sigma \to \mathbb{R}$ is *sub-harmonic* if, for every point $p \in \Sigma$ and every small enough positive real number $r > 0$, and given any local holomorphic coordinate[30],

$$v(p) \leq \frac{1}{2\pi} \int v(p + re^{i\theta})\, d\theta. \tag{7}$$

In the case where v is C^2, one can easily check that v is sub-harmonic if and only if its Laplacian[31] is non-positive. If v is only continuous, its Laplacian is not defined in the usual sense, but is nevertheless defined as a distribution[32]. This allows to generalize the preceding fact: a continuous function v is sub-harmonic if and only if its Laplacian is positive in a distributional sense, i.e., if it is a positive measure. Now let us recall that, when v is seen as an electrostatic potential, the Laplacian of v corresponds to the electric charge distribution which gives rise to the potential v. Hence, a potential $v : \Sigma \to \mathbb{R}$ is sub-harmonic if and only if the corresponding electric charge distribution is made only of negative charges.

Poincaré describes everything under the electrostatic analogy: he considers an electric charge distribution on A which is made of a positive point charge at p_0 and a distribution of negative charges on $A - \{p_0\}$; the electric potential corresponding to this electric charge distribution – which he calls u_0 – has a simple logarithmic singularity at p_0, and is sub-harmonic on $A - \{p_0\}$.

We now assume that the function $u_0 : A \to \mathbb{R}$ has been fixed (in such a way that it is continuous, non-negative, has a simple logarithmic singularity at p_0, and is sub-harmonic on $A - \{p_0\}$)[33]. Recall that a sequence of discs $(D_{i_k})_{k \geq 0}$ has been chosen (see Subsection 6.2), and that the sequence of function $(u_n)_{n \geq 0}$ is defined

[30] One needs a local holomorphic coordinate in order to give a meaning to the notation $p + re^{i\theta}$. Nevertheless, if inequality (7) for one local holomorphic coordinate (for every p and every $r > 0$), then it also holds for any other local holomorphic coordinate.

[31] Actually, on a Riemann surface which is not a domain in the complex plane, there is no well-defined Laplace operator. But, the differential 2-form given by the formula $\frac{\partial^2 v}{\partial x \partial y} dx \wedge dy$ in every local holomorphic coordinate $z = x + iy$ is well defined. This differential 2-form is often denoted by $dd^c v$. I shall abuse the language by calling this differential 2-form "the Laplacian of v"

[32] More exactly, the differential 2-form $dd^c v$ is defined in the distributional sense; see the preceding footnote.

[33] Notice that it is easy to construct such a function u_0. One may, for example, pick a local holomorphic coordinate z defined on a neighbourhood of p_0 and such that $z(p_0) = 0$, and consider the function u_0 which is defined by $u_0(p) = -\log|z(p) - z(p_0)|$ on the disc $|z| \leq r$, and is constant (equal to $\log r$) outside this disc.

as follows : for every n, the function u_{n+1} is obtained from u_n by sweeping the electric charges outside D_{i_n}.

Fact 25. *Assume that the sequence $(u_n)_{n\geq 0}$ converges uniformly on every compact set of $A-\{p_0\}$ towards a function u. Then u is non-negative, harmonic on $A-\{p_0\}$, and has a simple logarithmic singularity at p_0.*

To prove this fact, let us first observe that the definition of the sequence (u_n) and the maximum principle ensure that the function u_n is non-negative for every n. The limit u will therefore be non-negative. Now, pick an integer ℓ, and consider the subsequence $(u_{n_k})_{k\geq 0}$ such that, for every k, the potential u_{n_k} is obtained from the potential $u_{n_k} - 1$ by sweeping the electric charges outside the disc D_ℓ (such a subsequence exists since we have assumed that each element of the family $\mathcal{D}$ appears infinitely many times in the sequence D_{i_n}). First assume that p_0 is not in D_ℓ. Then u_{n_k} is harmonic on D_ℓ for every k. Since u_{n_k} converges to u uniformly on D_ℓ, one deduces that u is harmonic on D_ℓ. If p_0 is in D_ℓ, the same arguments prove that u is harmonic on $D_\ell - \{p_0\}$ and has a simple harmonic singularity at p_0. Therefore, u is harmonic on $A - \{p_0\}$ and has a simple harmonic singularity at p_0. This concludes the proof of Fact 25.

Fact 26. *For every n, the function u_n is sub-harmonic on $A - \{p_0\}$ and has a simple logarithmic singularity at p_0. Moreover, $u_{n+1} \geq u_n$ for every n.*

The proof of this fact is done by induction over n. Let us assume that u_n is sub-harmonic on $A - \{p_0\}$ and has a simple logarithmic singularity at p_0. Recall that u_{n+1} is obtained from u_n by sweeping the electric charges outside the disc D_{i_n}. In particular, u_{n+1} is harmonic on D_{i_n} (with a simple logarithmic singularity at p_0 if p_0 is in D_{i_n}). By assumption, u_n is sub-harmonic on D_{i_n} with a simple logarithmic singularity at p_0 if p_0 is in D_{i_n}. Therefore, $u_n - u_{n+1}$ is sub-harmonic on D_{i_n}. Moreover, $u_n - u_{n+1}$ vanishes on ∂D_{i_n}. By the maximum principle (for sub-harmonic functions), it follows that $u_n \leq u_{n+1}$ on D_{i_n}. But u_{n+1} coincides with u_n outside D_{i_n}. Therefore, the inequality $u_{n+1} \geq u_n$ holds everywhere. The fact that u_{n+1} is sub-harmonic (with a logarithmic pole at p_0) can be checked by a direct calculation (using the inequality $u_n \leq u_{n+1}$); we prefer to present Poincaré's physical "proof" of this fact: to go from u_n to u_{n+1}, we have "swept the electric charges that where in the disc D_{i_n} to the boundary of this disc"; since there were only negative charges in $A - \{p_0\}$ before this operation (recall that this is equivalent to say that u_n is sub-harmonic on $A - \{p_0\}$), one still has only negative charges after this operation. Therefore u_{n+1} is sub-harmonic on $A-\{p_0\}$!

In the next step, Poincaré uses the so-called *Harnack's principle*:

Proposition 27 (Harnack's principle). *Let Ω be a simply connected open domain of a Riemann surface, and (u_k) be an increasing sequence of sub-harmonic functions on Ω. Then (u_k) converges uniformly on compact sets either to $+\infty$, or to a sub-harmonic function.*

A proof of Harnack's principle can be found, e.g., in [St-Gervais2010, Chapter XII]. Facts 25 and 26, combined with Harnack's principle, imply the following:

Proposition 28. *If there exists a point $p \in A - \{p_0\}$ such that the sequence of real numbers $(u_n(p))_{n\geq 0}$ is bounded, then the sequence of functions $(u_n)_{n\geq 0}$ converges uniformly on compact sets to a function $u : A \to \mathbb{R}$ which is positive, continuous, harmonic on $A - \{p_0\}$, and has a simple logarithmic singularity at p_0 (in particular, u is a generalized Green function).*

So, it remains to find a point $p \in A - \{p_0\}$ such that the sequence $(u_n(p))_{n\geq 0}$ is bounded. The key argument is the following:

Proposition 29. *The integral over A of the Laplacian*[34] *of u_n does not depend on n.*

Poincaré seems to consider this proposition as physical evidence: he does not provide a proof for it. Recall that, when u_n is regarded as an electrostatic potential, the Laplacian of u_n is the electric charge distribution that gives rise to this potential. Hence, Proposition 29 just means that the total electric charge does not change during the sweeping process. This is indeed quite a "physical evidence": to go from u_n to u_{n+1}, one makes the disc D_{i_n} become an electric conductor; the electric charges which were in D_{i_n} are swept to the boundary of D_{i_n}, but the total electric charge in A does not change! For a more mathematical proof, the reader is referred to [St-Gervais2010, Proposition XIII.2.4].

Proposition 29 will be used to prove that the integral of u_n over a certain curve is bounded independently of n. For $s > 0$, we denote by $\mathbb{D}(0, s)$ the open disc of radius s centered at 0 in the complex plane. The following technical lemma easily follows from Green's formula (see [St-Gervais2010, Lemma XIII.2.5]):

Lemma 30. *Let r, r' be two real numbers such that $0 < r < r' < 1$. If $u : \mathbb{D}(0,1) \to \mathbb{R}$ is a continuous sub-harmonic function which vanishes on the disc $\mathbb{D}(0, r)$, then*

$$\frac{1}{2\pi}\int_0^{2\pi} u(r'e^{i\theta})d\theta \leq \log\frac{r'}{r}\int_{\mathbb{D}(0,r')}\Delta u. \tag{8}$$

To apply this lemma to our situation, let us recall how the Riemann surface A was constructed. We started with a simply connected non-compact Riemann surface S. We chose a local holomorphic chart z mapping an open set of S onto a neighbourhood of the origin in $\mathbb{C}$. Then we chose a positive real number r such that the closed disc $\overline{\mathbb{D}}(0, r)$ is contained in the range of z, and we set $A := S \setminus z^{-1}(\overline{\mathbb{D}}(0, r))$. Of course, we may assume that $r < 1$.

Now, let us choose a positive real number r' such that $1 > r' > r$, and such that the closed disc $\overline{\mathbb{D}}(0, r')$ is still contained in the range of z. Let us denote $D := z^{-1}(D(0, r))$ and $D' := \phi^{-1}(D(0, r'))$. For every n, let $\overline{u}_n : S \to \mathbb{R}$ be the function that coincides with u_n on $A = S \setminus \overline{D}$, and vanishes on $\overline{D}$. Lemma 30

[34]See footnote 31

shows that, for every $n \geq 0$, we have

$$\int_{\partial D'} u_n = \int_{\partial D'} \overline{u}_n \leq \log \frac{r'}{r} \int_{D'} \Delta \overline{u}_n = \log \frac{r'}{r} \int_{D' \setminus \overline{D}} \Delta u_n \leq \log \frac{r'}{r} \int_A \Delta u_n \quad (9)$$

(where Δu_n should be understood in the sense of footnote 31).

Inequality (9) and Proposition 29 show that the integral of u_n along the closed curve $\partial D'$ is bounded from above independently of n. One easily deduces the existence of a point $p \in \partial D'$ such that the sequence $(u_n(p))_{n \geq 0}$ is bounded from above. According to Fact 28, this implies that u_n converges, uniformly on compact sets of $A - \{p_0\}$, to a generalized Green function. By Proposition 23, this implies that the Riemann surface S is biholomorphic to the unit disc, and therefore concludes the proof of Theorem 20.

6.4. A simplified proof by Koebe

Koebe's first proof of Theorem 20 had been published a two months before the publication of Poincaré's memoir. As he reads Poincaré's proof, Koebe understands that this proof can be simplified quite radically. He immediately writes a short article in which he proposes a kind of shortcut to show the existence of a generalized Green function on the annulus A ([Koebe1907b]).

In fact, Koebe did realize that Poincaré's proof contains a truly new argument: the fact that the integral over A of the Laplacian of u_n does not depend on n (Proposition 29; recall that Poincaré just says that this fact expresses the global electric charge conservation). Koebe keeps this crucial argument, but replaces Poincaré's *sweeping process* by a much simpler process. He considers an exhaustion of A by an increasing sequence of compact annuli $(A_n)_{n \geq 0}$, and fixes a point p_0 in A_0. For each n, he knows that the annulus A_n admits a Green function u_n with its pole at p_0. It remains to show that the sequence (u_n) converges to a generalized Green function. For this purpose, Koebe adapts Poincaré's "electric charge conservation argument" and gets a upper bound (which does not depend on n) for the integral of u_n over a circle close to the boundary of A. He deduces the existence of a point p such that the sequence $(u_n(p))_{p \geq 0}$ is bounded and completes the proof using Harnack's principle.

Just as Poincaré, Koebe ignores all the issues concerning the smoothness of the functions he considers. Koebe's proof is nevertheless valid: indeed, one can easily verify (using completely elementary arguments) that the functions considered by Koebe are actually smooth. This is a considerable advantage of Koebe's proof (recall that one needs to use distribution theory to make Poincaré's arguments rigorous). But the beautiful electrostatic analogy invented by Poincaré has suddenly completely disappeared...

References

[Berger1971] M.S. Berger, Riemannian structures of prescribed Gaussian curvature for compact 2-manifolds. J. Differential Geometry **5**, 325–332 (1971).

[Fuchs1866] L. Fuchs, Zur Theorie der linearen Differentialgleichungen mit veränderlichen Coeffizienten. J. Reine Angew. Math **66**, 121–160 (1866).

[Fuchs1880] ———, Über die Functionen, welche durch Umkehrung der Integrale von Lösungen der linearen Differentialgleichungen entstehen. Nachr. Ges. Wiss. Göttingen, 445–453 (1880).

[Jost2002] J. Jost, *Compact Riemann surfaces, An Introduction to Contemporary Mathematics, Second edition.* Springer, New York, 2002.

[Klein1881] F. Klein, Über die conforme Abbildung von Flächen. Math. Ann. **19**, 159–160 (1881).

[Klein1882a] ———, Über eindeutige Functionen mit linearen Transformationen in sich. Math. Ann. **19**, 565–569 (1882).

[Klein1882b] ———, Über eindeutige Functionen mit linearen Transformationen in sich. Math. Ann. **20**, 49–51 (1882).

[Klein1882c] ———, *Über Riemanns Theorie der algebraischen Funktionen und ihrer Integrale.* Teubner, Leipzig, 1882. Traduction anglaise (édition révisée): *On Riemann's Theory of Algebraic Functions and their Integrals*, Dover, Mineola (NY), 2003.

[Klein1882d] ———, Neue Beiträge zur Riemann'schen Functionentheorie. Math. Ann **21**, 141–218 (1882).

[Klein1921a] ———, *Gesammelte Mathematische Abhandlungen.* 3 volumes, Springer, Berlin, 1921–1923.

[KleinPoincaré1923] F. Klein, H. Poincaré, La correspondance d'Henri Poincaré et de Félix Klein. Acta Math. **39**, 94–132 (1923).

[Koebe1907a] P. Koebe, Über die Uniformisierung beliebiger analytischer Kurven. Nachr. Ges. Wiss. Göttingen, 191–210 (1907).

[Koebe1907b] ———, Über die Uniformisierung beliebiger analytischer Kurven (zweite Mitteilung). Nachr. Ges. Wiss. Göttingen, 633–649 (1907).

[Osgood1998] B. Osgood, Old and new on the Schwarz derivative. In: Quasiconformal mappings and analysis (Ann Arbor, 1995), Springer, New York, 1998, pp. 275–308.

[Osgood1900] W. Osgood, On the existence of the Green's function for the most general simply connected plane region. Trans. Amer. Math. Soc. **1**, 310–314 (1900).

[Picard1890] É. Picard, Mémoire sur la théorie des équations aux dérivées partielles et la méthode des approximations successives. J. Math. Pures Appl. (4e série) **6**, 145–210 (1890).

[Picard1893a] ———, De l'équation $\Delta u = ke^u$ sur une surface de Riemann fermée. J. Math. Pures Appl. (4e série) **9**, 273–292 (1893).

[Picard1893b] ———, Sur l'équation $\Delta u = e^u$. C.R. Ac. Sci. Paris **116**, 1015–1017 (1893).

[Picard1893c] ———, Sur une équation aux dérivées partielles. C.R. Ac. Sci. **116**, 454–456 (1893).

[Picard1898] ———, De l'équation $\Delta u = e^u$. J. de Math. Pures et Appl. (5e série) **4**, 313–316 (1898).

[Picard1900] ———, De l'intégration de l'équation $\Delta u = e^u$ sur une surface de Riemann fermée. Bull. Sci. Math. **24**, 196–210 (1900). = J. Reine Angew. Math **130**, 243–258 (1905).

[Picard1931] ———, *Quelques Applications Analytiques de la Théorie des Courbes et des Surfaces Algébriques.* Gauthier-Villars, Paris, 1931.

[Poincaré1881a] H. Poincaré, Sur les fonctions fuchsiennes. C.R. Acad. Sci. Paris **92**, 333–335 (1881).

[Poincaré1881b] ———, Sur les fonctions fuchsiennes. C.R. Acad. Sci. Paris **92**, 395–396 (1881).

[Poincaré1881c] ———, Sur les fonctions fuchsiennes. C.R. Acad. Sci. Paris **92**, 859–861 (1881).

[Poincaré1881d] ———, Sur les fonctions fuchsiennes. C.R. Acad. Sci. Paris **92**, 1198–1200 (1881).

[Poincaré1881e] ———, Sur les fonctions fuchsiennes. C.R. Acad. Sci. Paris **93**, 301–303 (1881).

[Poincaré1882a] ———, Mémoire sur les fonctions fuchsiennes. Acta Math. **1**, pp. 193–294 (1882).

[Poincaré1882b] ———, Théorie des groupes fuchsiens. Acta Math. **1**, 1–62 (1882).

[Poincaré1883] ———, Sur un théorème de la théorie générale des fonctions. Bull. Soc. Math. France **11**, 112–125 (1883).

[Poincaré1884] ———, Sur les groupes des équations linéaires. Acta Math. **4**, 201–312 (1884).

[Poincaré1898] ———, Les fonctions fuchsiennes et l'équation $\Delta u = e^u$. J. Math. Pures Appl. (5e série) **4**, 137–230 (1898).

[Poincaré1907] ———, Sur l'uniformisation des fonctions analytiques. Acta Math. **31**, 1–64 (1907).

[Poincaré1908] ———, L'invention mathématique. Conférence à l'Institut de Psychologie, 23 mai 1908. Enseignement Math. **10**, 357–371 (1908).

[Rowe1989] D. Rowe, Klein, Hilbert, and the Göttingen mathematical tradition. Osiris (2) 5, 186–213 (1989).

[St-Gervais2010] H.-P. de Saint-Gervais, *Uniformisation des Surfaces de Riemann. Retour sur un Théorème Centenaire.* ENS Éditions, Lyon, 2010.

[Schwarz1870] H. Schwarz, Über einen Grenzübergang durch alternierendes Verfahren. Wolf J. Vierteljahresschrift der Naturforschenden Gesellschaft in Zürich **XV**, 272–286 (1870). = *Gesammelte mathematische Abhandlungen* (2*ter Band*). pp. 133–143, Springer, Berlin, 1890.

[Weyl1913] H. Weyl, *Die Idee der Riemannschen Fläche.* Teubner, Leipzig, 1913. Reprint of the 1913 German original, Teubner, Stuttgart, 1997.

François Béguin
LAGA
Institut Galilée
Université Paris Nord
99 avenue Jean-Baptiste Clément
F-93430 Villetaneuse, France
e-mail: beguin@math.univ-paris13.fr

Henri Poincaré, 1912–2012, 231–233

Poincaré Seminar 2012

Harmony and Chaos

On the Figure of Henri Poincaré

A film by Philippe Worms
Produced by Vie des Hauts Production (Besançon – France)

Nicolas Bergeron, Thierry Dauxois, Étienne Ghys, Tadashi Tokieda, Alberto Verjovsky Sola and Cédric Villani meet in a house they rented for the occasion. Several days of work and reflection, far from everything, together. It doesn't happen so often and they can barely hide their joy.

Étienne, Cédric and the others are mathematicians, physicists ... There is the exuberant one, the taciturn, the cautious, the optimist, the enthusiast and the hesitant. All came to remember the one who, at a moment of their life, has been an inspirer, a trigger, an accomplice: Henri Poincaré.

In this enclosed, quiet place, where every detail makes it a quirky atmosphere, the discussions, the exchanges between the characters form the daily pattern of the small group.

In the course of a stroll in the park, while cooking, upon a reading session in the living room or a pool game, they start off conversations, put their ideas to the test, reveal the stakes.

As we move forward, we explore several fields of mathematics and physics, through their history and in relation to their topicality.

But to me, as for the characters of this film, this encounter is above all a way to tell, to display the most intimately the emotion of knowledge, and to bring it closer to a broad public.

Throughout my work I made every effort to shed light on the intense relationship between the scientists and an often elusive seeking of truth, both a source of emotion and an always renewed driving force. Poincaré expressed it like this: *"Seeking the truth should be the aim of our activity, it is the only worthy end. If we want to gradually free men from material concerns, it is so they can use their newly reacquired freedom to study and gaze at the truth."*

When I was preparing the film, Tadashi suggested one day that we *"start using our child's eyes again"*, and Nicolas that we seek what *"would make sense for us, curious visitors."*

There lies the spirit of the film, in this will to relay an emotion, to progress in our quest for something which makes sense for us curious visitors, fascinated, as mankind has always been, by the starry universe.

Main topics

1. Truth seeking
Everybody talks about the importance Poincaré had in their lives and how difficult it is to make a film on such a versatile and impressive character.

2. Trees and non-Euclidean geometry
In the park, a dead tree. Alberto and Etienne talk about trees and draw strange triangles. Later, after dinner, Nicolas, Etienne and Tadashi expose the omnipresence of non-Euclidean geometry, the importance of geometry in the way we see the world.

3. Rotating fluid masses, the shape of Earth and bifurcations
From a text by Poincaré, "Man has always been fascinated by the starry universe..." (introduction of *Hypothèses cosmogoniques*), the question of harmony and chaos is broached a first time, the problem of rotating fluid masses, of the shape of the Earth and the bifurcation concept, introduced by Poincaré and which has become omnipresent in 20th century's physics.

4. A mathematical world
Drinks before dinner in the park. It is the occasion to read Poincaré's famous text on "invention in mathematics". The occasion to near the heart of the work of mathematicians: do we create or do we discover mathematics?

5. Topology, Poincaré conjecture, Tadashi's paper clips and Borromean rings
As Tadashi is making dinner, Nicolas and Alberto explain to us rudiments of topology with melons, buoys and shoelaces. Later that evening, by the fireplace, Tadashi shows us one of his "tricks": with paperclips and a strip of paper, he displays Borromean rings.

6. Chaotic systems, the theory of limit cycles, the butterfly effect, harmony and chaos
It is stormy outside and everyone gathers around the pool game for an experiment on chaotic systems. To bring contrast, Thierry and Tadashi then present the limit cycles. Étienne and Cédric conclude the film: "What is the most common in Nature? Chaos or Harmony?"

Philippe Worms
Vie des Hauts Production
15 c Chemin des Essarts
F-25000 Besançon, France

e-mail: philippe.worms@wanadoo.fr

Copies of the film are available at `vdh.prod@wanadoo.fr`
or at `http://videotheque.cnrs.fr/doc=3737`.

MIX
Papier aus verantwortungsvollen Quellen
Paper from responsible sources
FSC® C105338

If you have any concerns about our products,
you can contact us on
ProductSafety@springernature.com

In case Publisher is established outside the EU,
the EU authorized representative is:
Springer Nature Customer Service Center GmbH
Europaplatz 3, 69115 Heidelberg, Germany

Printed by Libri Plureos GmbH
in Hamburg, Germany